Friction Stir Welding of Dissimilar Alloys and Materials

Friction Stir Welding of Dissimilar Alloys and Materials

Nilesh Kumar
Department of Materials Science and Engineering,
University of North Texas, Denton, TX, USA

Wei Yuan
Research & Development Division, Hitachi America, Ltd.,
Farmington Hills, MI, USA

Rajiv S. Mishra
Department of Materials Science and Engineering,
University of North Texas, Denton, TX, USA

AMSTERDAM • BOSTON • HEIDELBERG • LONDON
NEW YORK • OXFORD • PARIS • SAN DIEGO
SAN FRANCISCO • SINGAPORE • SYDNEY • TOKYO
Butterworth-Heinemann is an imprint of Elsevier

Butterworth-Heinemann is an imprint of Elsevier
The Boulevard, Langford Lane, Kidlington, Oxford OX51GB, UK
225 Wyman Street, Waltham, MA 02451, USA

Notices
Knowledge and best practice in this field are constantly changing. As new research and experience broaden our understanding, changes in research methods or professional practices, may become necessary.

Practitioners and researchers must always rely on their own experience and knowledge in evaluating and using any information or methods described herein. In using such information or methods they should be mindful of their own safety and the safety of others, including parties for whom they have a professional responsibility.

To the fullest extent of the law, neither the Publisher nor the authors, contributors, or editors, assume any liability for any injury and/or damage to persons or property as a matter of products liability, negligence or otherwise, or from any use or operation of any methods, products, instructions, or ideas contained in the material herein.

ISBN: 978-0-12-802418-8

Library of Congress Cataloging-in-Publication Data
A catalog record for this book is available from the Library of Congress

British Library Cataloguing-in-Publication Data
A catalogue record for this book is available from the British Library

For Information on all Butterworth-Heinemann publications
visit our website at http://store.elsevier.com/

This book has been manufactured using Print On Demand technology.

CONTENTS

Preface to This Volume of Friction Stir Welding and Processing Book Series

This is the fourth volume in the recently launched short book series on friction stir welding and processing. As highlighted in the preface of the first book, the intention of this book series is to serve engineers and researchers engaged in advanced and innovative manufacturing techniques. Friction stir welding was invented more than 20 years back as a solid state joining technique. In this period, friction stir welding has found a wide range of applications in joining of aluminum alloys. Although the fundamentals have not kept pace in all aspects, there is a tremendous wealth of information in the large volume of papers published in journals and proceedings. Recent publications of several books and review articles have furthered the dissemination of information.

This book is focused on joining of dissimilar alloys and materials, an area that is getting a lot of attention recently; and friction stir welding promises to be a breakthrough technique for this as well. The promise of friction stir welding for such joints lies in its ability to minimize the extent of intermetallic formation in dissimilar metals. The change in the flow behavior brings in additional challenges as well. There are early successful examples of implementation of dissimilar metal joining and hopefully this book will provide confidence to designers and engineers to consider friction stir welding for a wider range of dissimilar alloy and dissimilar metal joining. It will also serve as a resource for researchers dealing with various challenges in joining of dissimilar alloys and materials. As stated in the previous volume, this short book series on friction stir welding and processing will include books that advance both the science and technology.

Rajiv S. Mishra
Department of Materials Science and Engineering,
University of North Texas
February 16, 2015

CHAPTER 1

Introduction

Humans and materials have flocked together since the humans have roamed the earth. As a matter of fact, the influence of materials on human civilization has been so profound, our progress is sometimes described in terms of materials—stone age, copper age, bronze age, and iron age. Industrial revolution was a major turning point in the history of human civilization which propelled the development of new materials. New materials enabled building of stronger and cheaper artifacts used in a variety of situations such as ground, sea, and aerospace transportation-related applications. The twentieth century witnessed a phenomenal growth on the materials development front, and designers of engineering structure were presented with a monumental task of selecting an appropriate or a set of materials for a particular component. On the one hand the availability of a wide spectrum of materials allowed designers to be very creative with the design of any component, on the other it posed a new set of challenges in terms of integrating different types of materials in a single structure. Among many, the assembly of components made of materials widely differing in chemical, thermal, physical, and mechanical properties became a challenge. For the majority of dissimilar materials, mechanical fastening is an appropriate choice. But the demand on high-performance structures has shifted attention from mechanical joining such as riveting and bolting to welding. Although a great number of welding techniques have been developed so far to deal with different types of materials, the welding of dissimilar materials still remains a challenge.

1.1 EXAMPLES OF ENGINEERING SYSTEMS NEEDING DISSIMILAR JOINTS

The need for joining dissimilar materials often arises in industrial applications due to demand for a wide variety of materials to impart complex shape, different loading or performance conditions needed in different

Friction Stir Welding of Dissimilar Alloys and Materials. DOI: http://dx.doi.org/10.1016/B978-0-12-802418-8.00001-1

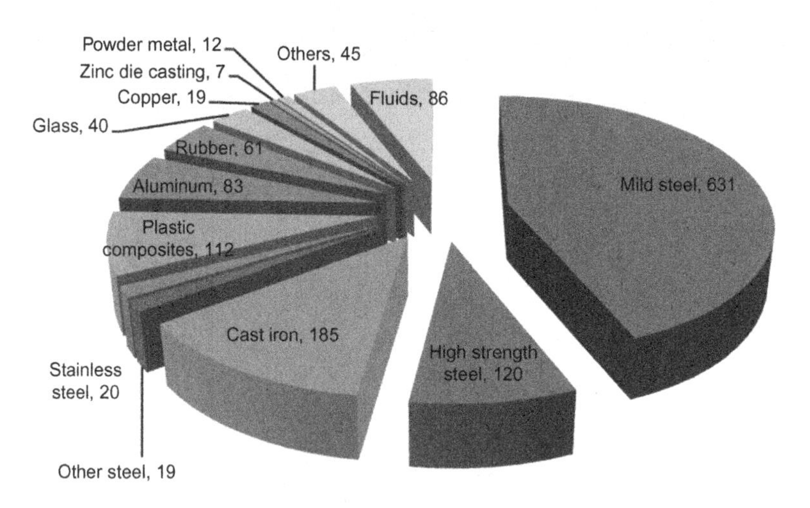

Figure 1.1 Material distribution of total vehicle curb weight in kilogram (Mayyas et al., 2012).

parts of the assembly such as high strength and corrosion resistance. Materials have been the backbone for industry, and advanced lightweight materials are essential especially for transportation industries to improve fuel economy while maintaining or improving safety and performance. Steels, owing to their attractive properties, recyclability, matured state of the art, and relatively low cost, have historically been the preferred choice for structural application in automotive industry. However, it is becoming clear that not a single material can fit all applications. The multi-material concept including a hybrid of light metals is now a trend for the automotive industry (Figure 1.1). With extra push toward the use of light materials, the fraction of light materials including polymer matrix composites is poised to increase in the near future. Traditional steel components can be replaced or partially replaced with lighter materials such as advanced high-strength steel, aluminum alloys and polymer.

One area where dissimilar material joint is essential in a structure is the fabrication of tailor-welded blank (Figure 1.2). A tailor-welded blank consists of joining sheets of different materials and/or the same material with different thicknesses, which is then submitted to a stamping process to form into the desired shape. Tailor-welded blanks are primarily used in the automotive industry and offer a significant potential on weight reduction for applications such as side frames, doors, pillars, and rails, because no reinforcement is required. The main

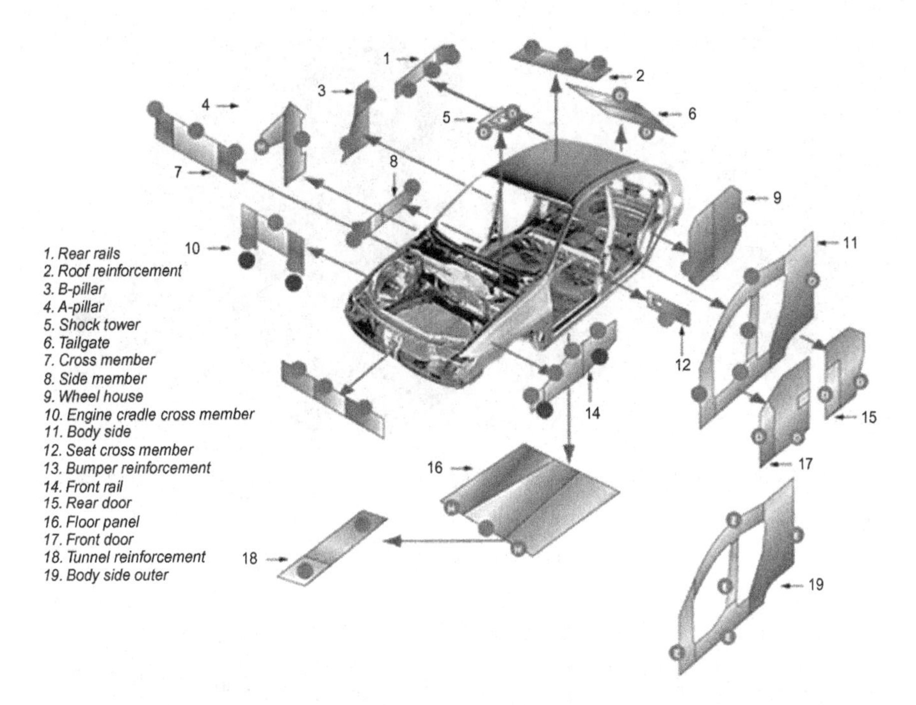

Figure 1.2 Tailor-welded blanks for automotive application (http://automotive.arcelormittal.com/tailoredblanks/TB_products/Applications, last accessed on 01.12.15).

advantage of a tailor-welded blank is that it allows the joining of multiple pieces to fabricate much larger components as well as proper distributions of weight and material properties in the final stamped part with a consequent reduction in weight and cost.

In addition to the body structure, there are also components and devices in an automobile consisting of dissimilar material joints, such as powertrain components. Figure 1.3 presents a turbocharger impeller for high-efficiency gas and diesel engines. The impeller is made of carbon steel and Inconel and welded by using electron beam. The dissimilar material assembly enables the lightweight design as well as superior performance.

Advanced materials, structures, and fabrication technologies are needed to enable the design and development of advanced future aircraft especially in airframe and propulsion systems. The high-performance materials such as titanium alloy and nickel-based superalloy, and adaptive materials such as piezoelectric ceramics, shape memory alloys, shape memory polymers, and carbon fibers, can only be

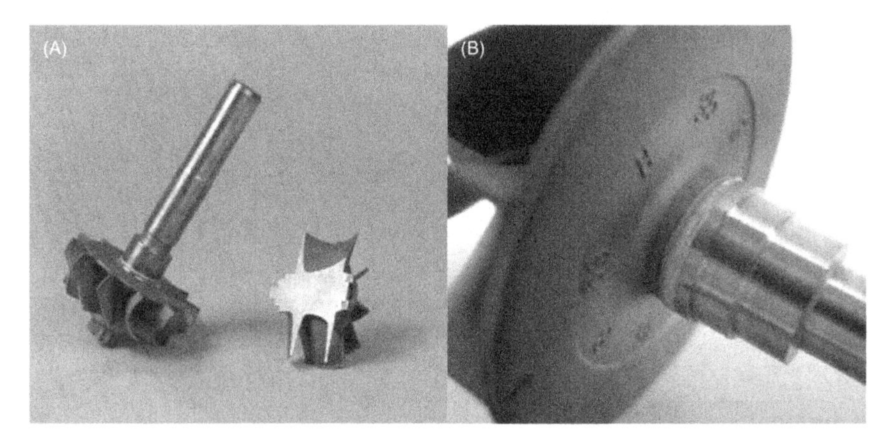

Figure 1.3 (A) Turbocharger impeller made of carbon steel (shaft) and Inconel (impeller). (B) Magnified view of the weld between shaft and impeller (http://www.ptreb.com/industries/automotive/turbocharger_impeller_welding/, accessed on 30.11.14).

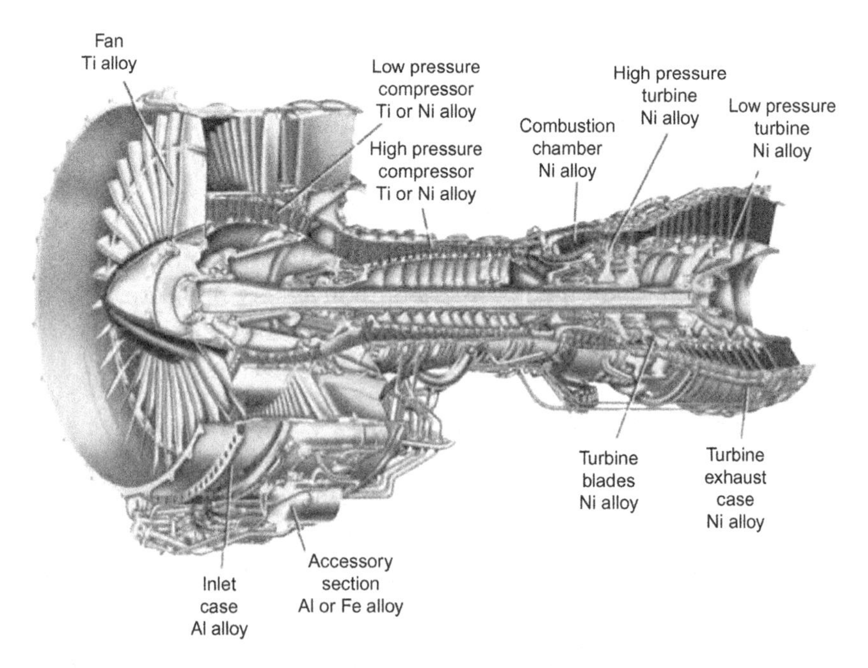

Figure 1.4 Aircraft engine with multifunctional materials (Campbell, 2006).

used where they are essential. To integrate these materials into airframe and/or aircraft engine structures, development of joining and integration technologies including metal to metal and metal to ceramic is critical (Figure 1.4).

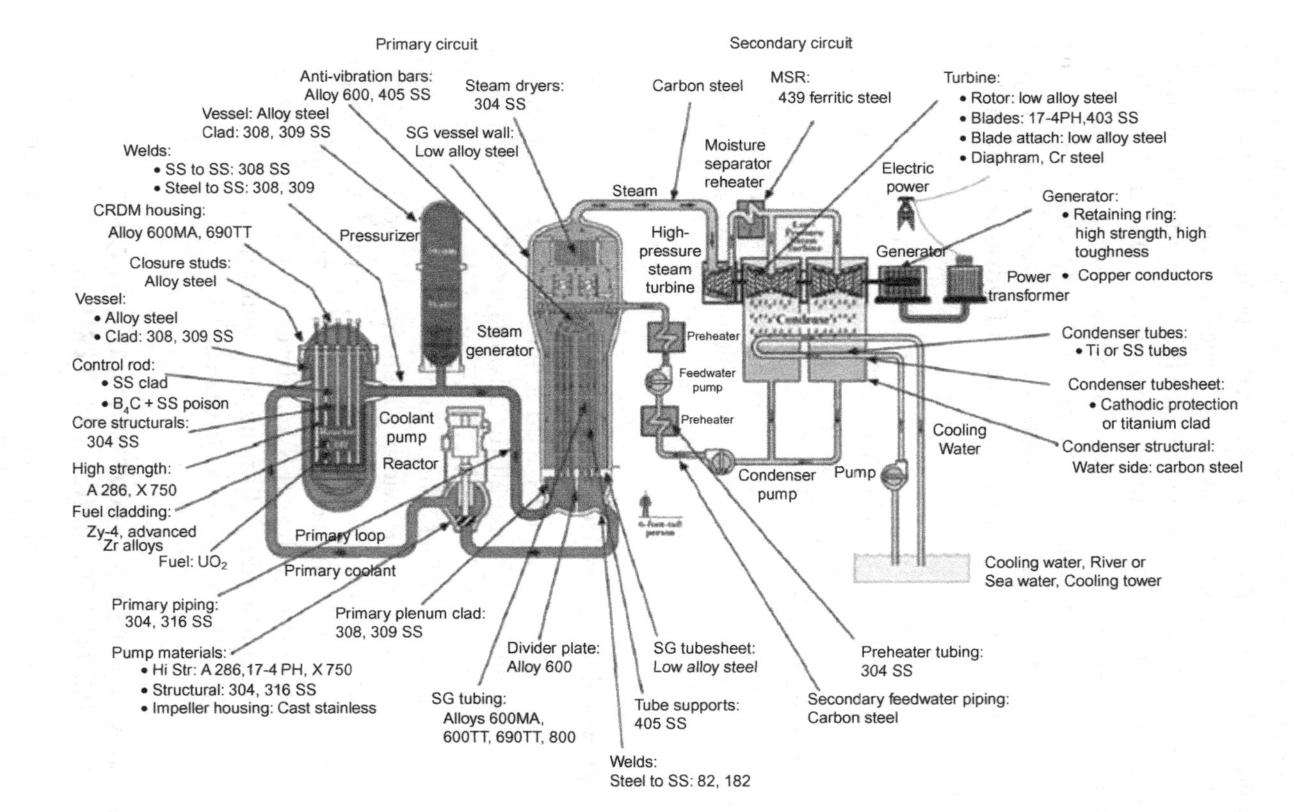

Figure 1.5 A schematic diagram showing different parts of a pressurized light-water nuclear reactor and materials used in the construction (Zinkle and Was, 2013).

Figure 1.5 shows a schematic of pressurized light-water nuclear reactor showing the use of an array of different high-temperature materials in the primary and secondary circuits including the reactor. It is appreciated that materials with better performance would be needed in reactors being designed for longer lifetime and superior capability. Similar need is being felt in the development of ultra-supercritical steam boilers expected to operate at 760°C and 35 MPa. The current design allows boilers to operate at 620°C and 20 MPa (Sridhar et al., 2011). Again, to meet increased expectation from the materials will require not only development of materials with increased performance but also development and use of advanced integration technologies.

Above examples taken from various industries show that a wide range of materials are needed for successful performance of engineering structures. A large number of components made of different materials are integrated to give rise to the final structure. A number of engineering solutions are available to assemble subsystems and choice of which depends on a number of factors including availability, cost, and performance expected from the system.

1.2 CONVENTIONAL JOINING TECHNIQUES

Figure 1.6 shows a few commonly used joining techniques for similar and dissimilar materials. Among all the advantages of welding include cheaper and faster integration time, flexibility in design, weight savings, higher structural stiffness, high joint efficiency, air and water tightness, and no limit on the width which can be welded together. Conventionally both fusion welding and solid-state welding techniques have been used to join dissimilar materials. Solid-state welding techniques include friction stir welding (FSW), ultrasonic welding, explosion welding, and diffusion welding. Brazing and soldering also have been tried to create joints between dissimilar welds.

1.3 DISADVANTAGES OF CONVENTIONAL WELDING TECHNIQUES FOR DISSIMILAR MATERIALS

Although the majority of the issues encountered in the dissimilar metal welding using fusion welding techniques are present in solid-state welded joints, it is less severe in solid-state welds. Some of the

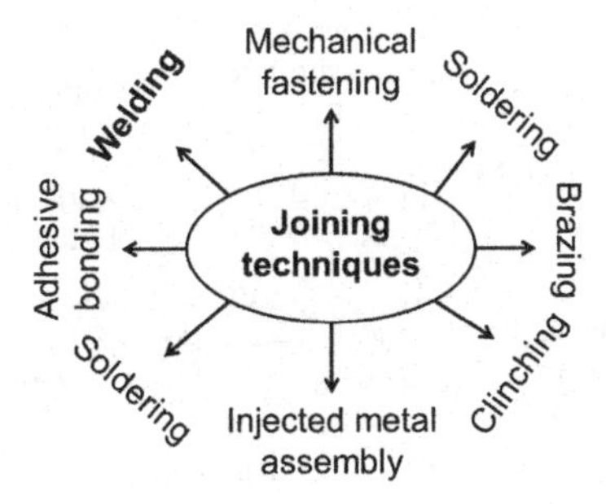

Figure 1.6 Various techniques used to join similar and dissimilar materials.

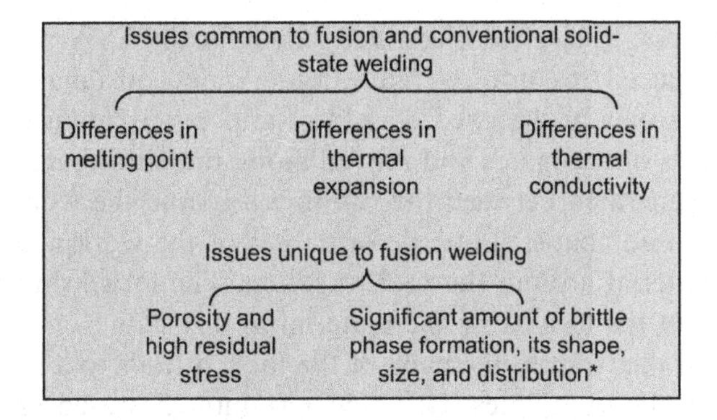

*Figure 1.7 A few issues commonly encountered in dissimilar metal welding using fusion and conventional fusion welding techniques (*actually formation of brittle intermetallic phase is also found in solid-state welding; however, it is less severe in solid-state welding than that in fusion welding).*

advantages of the solid-state weld over fusion welds with regard to similar welds still hold good during dissimilar material welding—for example, the absence of porosity and less distortion during solid-state welds. Due to high temperature during fusion welding compared to solid-state welding, most of the time the use of filler material results in a weld material where metallurgical characteristics, mechanical, and physical properties are totally different from individual materials used in dissimilar welds. Some of the issues faced during solid-state welding and fusion welding are depicted in Figure 1.7.

1.4 FRICTION STIR WELDING

Among all the solid-state welding techniques, FSW is a relatively new technique. FSW is a solid-state joining process invented in 1991 at The Welding Institute, UK (Thomas et al., 1995). It is a remarkably simple

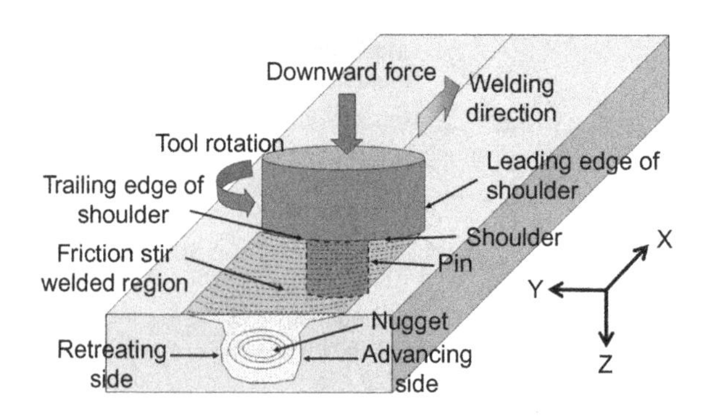

Figure 1.8 A schematic drawing of FSW in a butt joint configuration (Mishra and Ma, 2005).

welding process. Figure 1.8 illustrates schematically FSW processes for two plates placed in butt configuration. A nonconsumable rotating tool having specially designed shoulder, and pin is plunged into the abutting edges of the plates and moved along the parting line. The frictional heat generated between the rotating tool and the workpiece and the heat from adiabatic plastic deformation of the workpiece material cause the material around the tool to soften. The forward motion and the rotation of the tool cause the material in this state to move around the tool from the front to the back of the tool. It leads to a joint formation between the two plates. Different terminologies used in FSW are also labeled in the schematic shown in Figure 1.8. Most definitions are self-explanatory, but advancing side and retreating side definitions require a brief explanation. The side of the weld where the sense of tangential velocity of the rotating tool is parallel to the sense of the tool traverse is termed as advancing side, and if opposite, retreating side.

Note that, as the name suggests, the entire process of weld formation takes place below the melting point or solidus of the alloy. It leads to avoidance of most of the issues associated with fusion welding of materials. Additional key benefits of FSW as compared to fusion welding are summarized in Table 1.1.

Figure 1.9 illustrates various zones representing different microstructural states in the weldments as observed on the transverse cross-section of friction stir welds. The zone D represents dynamically recrystallized zone referred to as nugget. Thermo-mechanically affected zone (TMAZ), labeled here as region C, is the region of the weld which does not undergo complete recrystallization. Nugget and TMAZ regions experience plastic deformation at high temperature.

Table 1.1 Key Benefits of FSW (Mishra and Ma, 2005)

Metallurgical Benefits	Environmental Benefits	Energy Benefits
• Solid-phase process • Low distortion • Good dimensional stability and repeatability • No loss of alloying elements • Excellent mechanical properties in the joint area • Fine recrystallized microstructure • Absence of solidification cracking • Replace multiple parts joined by fasteners • Weld all aluminum alloys • Post-FSW formability	• No shielding gas required • Minimal surface cleaning required • Eliminate grinding wastes • Eliminate solvents required for degreasing • Consumable materials saving, such as rugs, filler wire, or any other gases • No harmful emissions • No radiant energy as in fusion welding; hence simple safety glasses enough	• Improved materials use (e.g., joining different thickness) allows reduction in weight • Only 2.5% of the energy needed for a laser weld • Decreased fuel consumption in lightweight aircraft, automotive, and ship applications

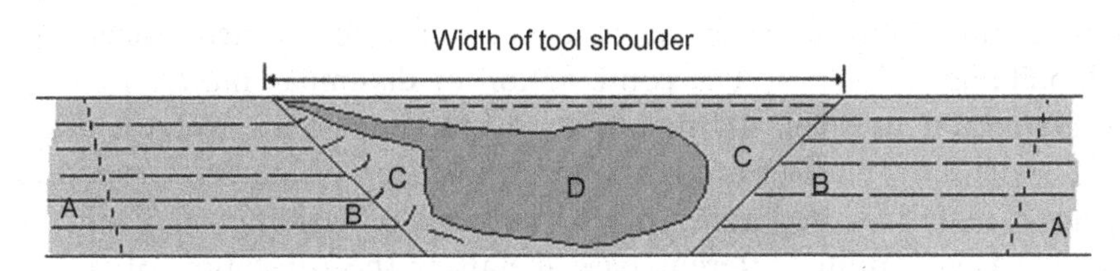

Figure 1.9 A schematic of transverse cross-section showing different zones of a friction stir weld. A, BM; B, HAZ; C, TMAZ; and D, Nugget (Mishra and Mahoney, 2007).

Heat-affected zone (HAZ) is the region of the weldments where only the influence of thermal excursion is present. Base material (BM) is the region which has the same set of properties as in as-received condition. The microstructural states of these zones are heavily dependent on FSW parameters.

The advent of FSW has completely revolutionized the field of welding. FSW is considered to be the most significant development in metal joining in decades. The high-strength aluminum alloys, such as 2XXX and 7XXX series aluminum alloys, are classified as "nonweldable" by fusion techniques. So, when the FSW was invented in 1991, it opened up new opportunities to weld high-strength aluminum alloys. Table 1.2 shows an example of strength levels achieved in initial studies of FSW of AA2024 and AA7075. Subsequently, other attributes of FSW, like defect-free welds, lower residual stresses, and lower distortion, led to numerous implementations using lower strength aluminum alloys.

Table 1.2 An Example of Results for AA2024Al and AA7075Al Alloys That Led to Excitement for Implementation of FSW (Mishra et al., 2014)

Base Alloy and Temper	Parent Material	Gas-Shielded arc Welded Butt Joint		FSW	
	Tensile Strength (MPa)	Tensile Strength (MPa)	% of Parent	Tensile Strength (MPa)	% of Parent
2024-T3	485	Nonweldable	–	432	89
7075-T6	585	Nonweldable	–	468	80

1.5 APPLICATIONS OF FRICTION STIR WELDED DISSIMILAR MATERIALS

Joining of dissimilar materials is becoming increasingly important as engineers strive for reduced weight and improved performance from engineering structures. FSW has already been adopted extensively for joining aluminum alloys in automotive, rail, aircraft, aerospace, and shipbuilding industries. The combination of dissimilar materials, such as aluminum to steel, aluminum to magnesium, and steel to nickel base superalloy, enables an optimum exploitation of the best properties of both materials. A barrier placed in front of welding of dissimilar materials with quite different base metals is the formation of brittle intermetallic compounds, which diminishes the strength and integrity of a structure. Recent efforts on reducing such deleterious compounds by using FSW have led to the implementation and mass production of dissimilar materials structures for industrial applications.

The progress made in welding lightweight materials, such as aluminum and magnesium alloys, make the mass production of light transportation systems possible and hence a significant reduction in fuel consumption. FSW has been adopted by the automotive industry for more than a decade to join aluminum alloys. Recently, Honda Motor Corporation has implemented FSW to join dissimilar aluminum alloy and steel in an automobile front structural component in production vehicle Honda Accord. The front subframe which carries the engine and some suspension components, is made of die cast aluminum and press formed steel halves. FSW was applied to weld the aluminum to the steel in a lap configuration at various locations as indicated by short stitches. Honda claimed that the total body weight is reduced by 25% compared to the conventional steel subframe with reduced electricity consumption by approximately 50% (http://world.honda.com/news/2012/4120906Weld-Together-Steel-Aluminum/).

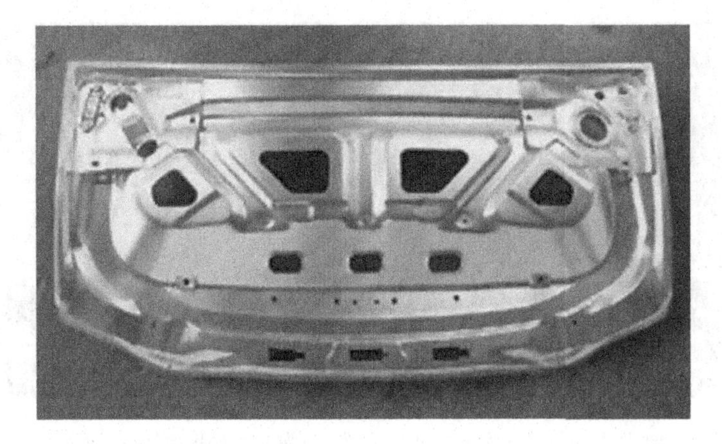

Figure 1.10 FSW to join the aluminum deck lid to galvanized steel brackets by Mazda (Mishra and Mahoney, 2007).

Mazda Motor Corporation has developed direct friction stir spot joining technology to weld aluminum alloy and steel, and applied it to join the trunk lid of the Mazda MX-5. Figure 1.10 shows Mazda's dissimilar friction stir welded deck lid with aluminum sheet to galvanized steel brackets. In addition to the prominent weight reduction, Mazda also claims that this technology improves the potential of coupling aluminum parts to steel in vehicle bodies and helps lower the costs of production.

Tailor blanks with various dissimilar materials combinations, such as different aluminum alloys, aluminum and magnesium alloys, aluminum and steel, have been researched via FSW for both automotive and aircraft applications. Figure 1.11 shows dissimilar tailor-welded blanks of 1 mm thick sheets of AA 5182 and AA 6016 aluminum alloys. The welded blanks were further formed by deep drawing cylindrical cups. Although the mechanical behaviors are different between two aluminum alloys, the weld line remained straight and aligned at the middle of the cups after deep drawing.

FSW of dissimilar metals has also been used in other sectors, such as the health industry. Figure 1.12 shows a vacuum-tight component in X-ray equipment of Siemens Medical Solutions. The component is fabricated by FSW domed aluminum sheets to flat stainless steel sheets, manufactured by Riftech GmbH in Greesthacht, Germany. A noticeable cost reduction by approximately 20% has been achieved due to significantly reduced defective welds by using FSW process.

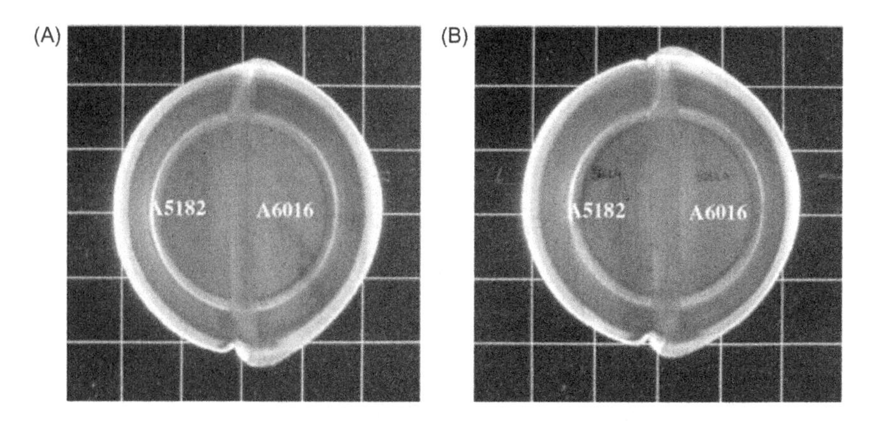

Figure 1.11 Tailor-welded blanks of dissimilar aluminum alloys welded by FSW and deep drawing under various blank-holder forces (force increases from left to right) (Leitão et al., 2009).

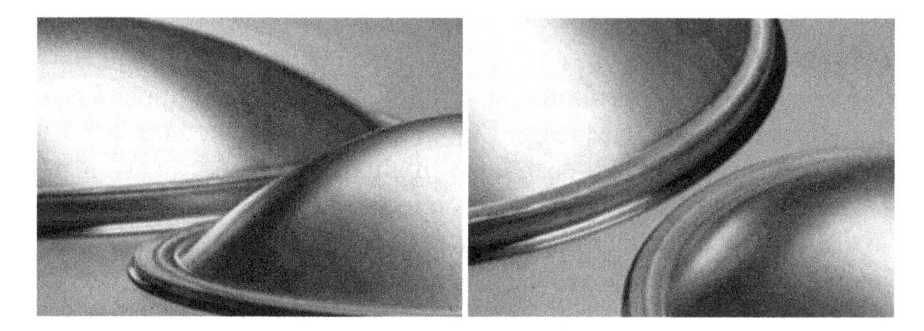

Figure 1.12 Friction stir welded high vacuum-tight aluminum-stainless steel joints in components for X-ray equipment (http://www.riftec.de/en/infopoint/news/industrial-friction-stir-welding-not-only-for-aluminium-materials/, last accessed on 01.12.15).

REFERENCES

Campbell, F.C., 2006. Manufacturing technology for aerospace structural materials. Elsevier Science (Chapter 6).

Leitão, C., Emílio, B., Chaparro, B.M., Rodrigues, D.M., 2009. Formability of similar and dissimilar friction stir welded AA 5182-H111 and AA 6016-T4 tailored blanks. Mater. Des. 30, 3235–3242.

Mayyas, A., Qattawi, A., Omar, M., Shana, D., 2012. Design for sustainability in automotive industry: A comprehensive review. Renew. Sust. Energ. Rev. 16, 1845–1862.

Mishra, R.S., Ma, Z.Y., 2005. Friction stir welding and processing. Mater. Sci. Eng. R 50, 1–78.

Mishra, R.S., Mahoney, M.W., 2007. Introduction, Friction Stir Welding and Processing. ASM International, Materials Park, OH. (Chapter 1, Chapter 12).

Mishra, R.S., De, P.S., Kumar, N., 2014. Friction Stir Welding and Processing: Science and Engineering. Springer, International Publishing Switzerland.

Sridhar, S., Rozzelle, P., Morreale, B., Alman, D., 2011. Materials challenges for advanced combustion and gasification fossil energy systems. Metall. Mater. Trans. A 42, 871–877.

Thomas, W.M., Nicholas, E.D., Needham, J.C., Murch, M.G., Smith, P.T., Dawes, C.J., 1995. Friction welding, PCT/GB92/02203, Patent No. 5,460,317.

Zinkle, S.J., Was, G.S., 2013. Materials challenges in nuclear energy. Acta Mater. 61, 735–758.

CHAPTER 2

A Framework for Friction Stir Welding of Dissimilar Alloys and Materials

2.1 ALLOY SYSTEMS

The ninth edition of Woldman's Engineering Alloys lists about 56,000 alloys for engineering applications (Frick, 2000). It highlights the daunting task materials scientists and engineers face during materials selection for various applications. Engineering materials can be broadly categorized into five families of materials—metals, ceramics, glasses, elastomers, and polymers. Materials from different families can also be combined together to give rise to what is called "hybrid" materials. For example, ceramic powders can be embedded in a metal matrix which gives rise to metal-matrix composite—a very important class of engineering materials falling under hybrid materials. Figure 2.1 schematically shows this classification.

With respect to dissimilar joining of engineering materials, those falling within a family can essentially be joined either with other classes of materials in the same family or with any materials from other families. For example, Al alloys in metals family can either be joined with Mg alloys present in the same family or materials in polymer family. It, therefore, opens up an astronomically large number of combinations of materials for dissimilar welding. However, the focus of this book is not to cover all combinations of dissimilar materials welding, but to select a few to showcase the benefits friction stir welding (FSW) brings in the joining of dissimilar materials and highlight issues, in general, associated with the welding of dissimilar materials.

Figure 2.2 very broadly classifies the dissimilar welding based on material combinations used during the welding. As per this convention, there exists, in general, three different combinations—Category I, under this category different alloys from the same class of metallic

Friction Stir Welding of Dissimilar Alloys and Materials. DOI: http://dx.doi.org/10.1016/B978-0-12-802418-8.00002-3

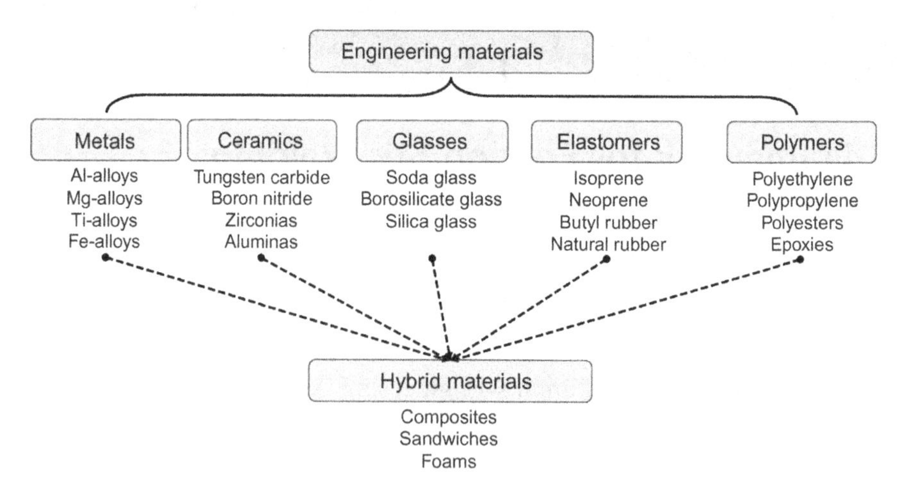

Figure 2.1 Family of materials. Adapted from Ashby (2005).

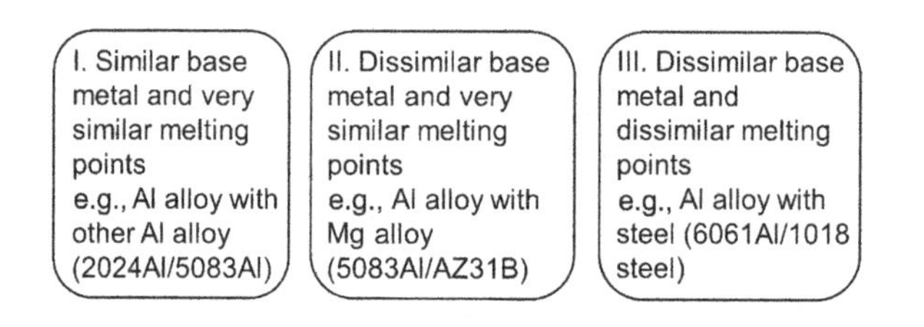

Figure 2.2 Classification of dissimilar materials welding based on type of metallic materials being welded.

materials are welded together; for example, welding of an aluminum alloy from one series (e.g., 2XXX series) with an aluminum alloy from another series (e.g., 7XXX series). Category II, materials that fall into this category belong to different classes of materials but melting temperatures (solidus temperature, to be precise) are not much different from each other; for example, joining of aluminum alloys with magnesium alloys. Category III, combinations of materials which not only belong to different classes of metallic materials but also have very different melting points fall in this category; for example, welding of aluminum alloys with steels. A detailed discussion of friction stir welded dissimilar materials will be carried out in Chapters 4 and 5 under two categories. Chapter 4 will discuss dissimilar alloys representing material combinations described in (I), and Chapter 5 will discuss dissimilar materials representing categories (II) and (III).

2.2 KEY SCIENTIFIC ISSUES IN THE FSW OF DISSIMILAR ALLOYS AND MATERIALS

Figure 2.3 is a conceptual diagram which has been used to illustrate key components of the process and also to highlight important friction stir scientific issues (Mishra, 2008). This figure can also be utilized to discuss key scientific issues pertaining to the FSW of dissimilar materials. Two key components in the FSW process are heat generation and material flow. Together, these determine the temperature history of the plate and give rise to the flow pattern and the resulting microstructure. Mathematical models and simulation tools pertaining to different aspects of any process are powerful means of developing a holistic understanding of the process. A robust understanding of the process calls for establishing a precise relationship between the input process variables (in this case, tool rotation rate, tool traverse speed, starting microstructure, etc.) and the output process variables (in this case, temperature, force, precipitates, etc.). However, even in similar welding (same alloy being welded to the same alloy) coupled thermo-mechanical nature of the FSW process and extremely complex materials flow during FSW have been major deterrence in developing a sound understanding about relationship among various process related input and output variables. Hence, our understanding at present, even

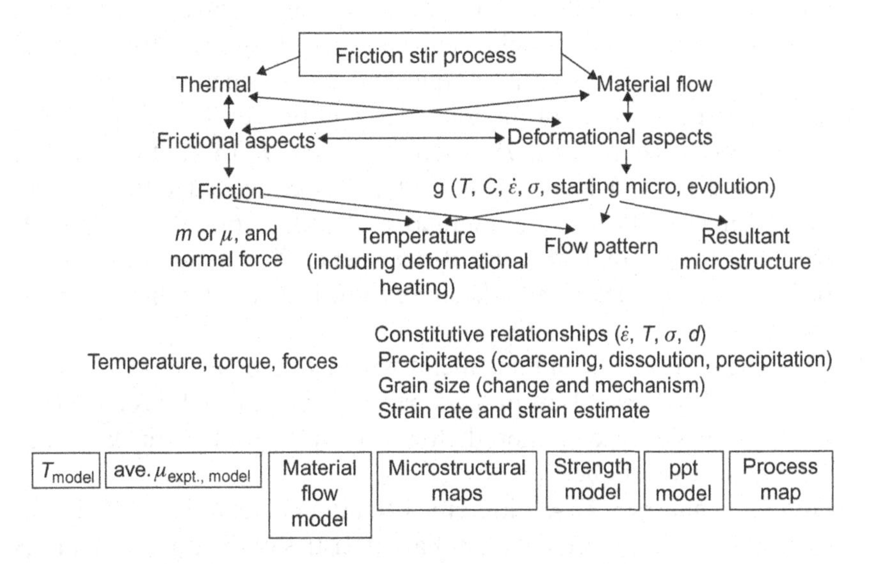

Figure 2.3 A schematic summarizing various aspects related to FSW process (Mishra, 2008).

after more than twenty years of FSW, about the correlation among various parameters is at best empirical in nature. It is not to say no effort has been made toward predictive analytical and simulation tools development. The presence of two materials with different physical and mechanical properties complicates the matter further.

A set of questions which are important during similar FSW are relevant even during dissimilar welding in addition to a few concerns pertinent only to dissimilar metal welding.

1. In the similar material welding, one has to worry about coefficient of friction (COF) just for one material and its dependence on thermal and deformational aspect of FSW process. However, the variation of COF in dissimilar welding is extremely challenging because the extent of mixing would determine the material in contact with the rotating tool. Hence, how exactly one should deal with choice of COF for modeling and simulation of dissimilar material welding is a tough question.
2. In the similar welding, thermal diffusivity ($\alpha = k/\rho C_p$) of both the plates is the same, and for all practical purposes the thermal profile in the workpieces can be considered symmetrical with respect to the weld centerline (small asymmetry in thermal profile might exist across weld centerline due to combined effect of tool rotation and tool traverse; advancing side (AS) and retreating side (RS)). However, in the dissimilar welding, α is expected to be different for both materials. It will lead to the development of an asymmetric thermal profile across weld centerline. Such asymmetric thermal profile is expected to affect material flow, microstructure, and mechanical properties of the joints. In addition to this, if one uses a simple relationship $F = \mu^* P$, where F, μ, and P are frictional force, COF, and normal load, respectively, it is clear that different frictional forces will exist on dissimilar materials during the welding. It will lead to different amount of heat generation at tool and workpiece interface. Hence, this becomes another variable in the asymmetry of the thermal profile.
3. As is the case for the similar metal welding, here also material flow is expected to depend on process parameters and tool design. What is the dependence of material flow on such variables in dissimilar welding is a valid question.
4. Different materials have different softening characteristics. This will have a major influence on flow characteristics of dissimilar materials

during welding. As an example, Figure 2.4 compares the flow characteristics of an aluminum alloy and a steel. During FSW of aluminum alloys in general a temperature of 450°C is commonly attained. At this temperature, AA6061-T6 and SS301 have widely differing yield strengths. Acknowledging the existence of asymmetric thermal profile and differences in materials softening characteristics, it is important to ask—for a given welding configuration how will tool geometries affect the material flow pattern? By changing the tool position with respect to weld centerline, thermal profiles can be affected and so can be the level of softening in each material. It leads to another question: whether there is any positional dependence of tool with respect to the weld centerline of the material flow.

5. Another equally important question is related to feasibility of welding materials widely differing in melting points by plunging FSW tool symmetrically about the joint line without causing melting of low-melting point material.
6. Once two different materials are welded together, it will result in a different chemistry of the welded zone. The resulting microstructure, especially amount, size, and form of second phases would heavily depend on resultant chemistry upon mixing of two different materials. This will affect the integrity of the joint.
7. Dissimilar materials have different coefficients of thermal expansion. How does it affect residual stresses and distortion of the resulting welded structure?

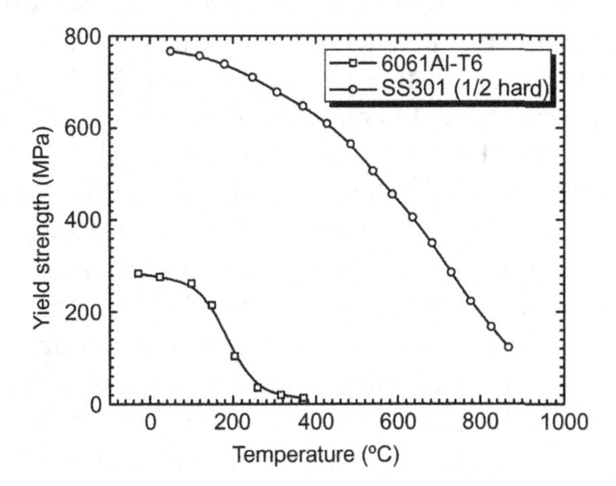

Figure 2.4 Softening characteristics of AA6061-T6 and SS301: yield strength versus temperature (Mishra et al., 2014).

Table 2.1 Thermophysical Properties of Some Pure Metals (Mills, 2002)

Metals	Density at 25°C (kg/m^3)	Coefficient of Thermal Expansion (1/K)/10^{-6} at 25°C	Heat Capacity (J/(K-kg)) at 25°C	Thermal Conductivity (W/(m K)) at 25°C	Melting Point (°C)
Al	2702	28	905	91	660.2
Co	8862	16.7 (25–900°C)	425	100	1495
Cu	8930	19	380	400	1084.6
Fe	7874	14.5 (20–900°C)	450	73	1538
Mg	1740	30	103	157	650
Ni	8900	17.3	426	90	1455
Si	2330	3.8	712	140	1414
Ti	4540	11	525	21	1668
Zn	7140	30	390	120	419.4

Table 2.1 lists thermophysical properties of a few pure metals. These metals have widely differing thermophysical properties, and consideration of these becomes important during dissimilar welding of metals or alloys based on these metals.

2.3 HEAT GENERATION AND TEMPERATURE DISTRIBUTION

The discussion and analytical estimation of heat generation at tool–workpiece interface in similar welding is relatively easier because of the presence of only one type of material in contact with the tool. However, things become quite complicated when it comes to heat generation and temperature distribution in dissimilar FSW. Two different materials, in general, have different thermal properties, COF, and softening characteristics (Figure 2.4 and Table 2.1). All these affect heat generation, temperature distribution in the workpieces, and material flow. Let us first consider heat generation aspect in FSW of dissimilar materials. In similar welding, the position of the tool rotation axis with respect to the weld centerline does not affect the heat generation. However, for dissimilar materials, it becomes important to consider the proportion of the tool surface contacting each material. The published literature on dissimilar Al alloy and steel suggests that if the tool is plunged mostly toward high-temperature material, the heat generation is so excessive that it causes melting of the low-melting temperature material. Hence a proper strategy needs to be adopted to avoid melting of low-melting temperature material.

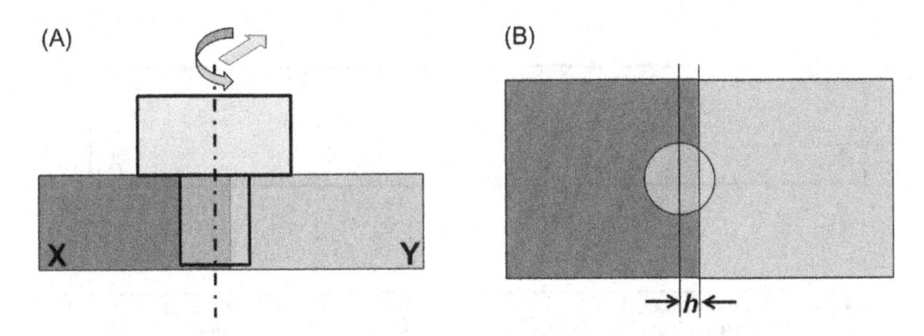

Figure 2.5 (A) A schematic of the transverse cross-section of the dissimilar weld showing asymmetrically located FSW tool, and (B) top view showing tool pin offset with respect to the weld centerline.

During FSW, a major component of total heat generated is due to frictional heating between the tool and the workpiece. In similar welding, the tool is in contact with the same type of material for the entire duration. However, as shown schematically in Figure 2.5, at any given time during each rotation, a part of the tool will be in contact with one type of material and the rest with another. Depending on the position of the tool with respect to the weld centerline and COF, the amount of heat generated in two materials is likely to vary significantly.

The next stage is thermal profile of the dissimilar metal weld. In the similar welding, for all practical purposes, the thermal profile can be considered symmetric with respect to the weld centerline. However, during dissimilar metal welding if significant difference in thermal diffusivity (defined as $k/(\rho C_p)$, where k = thermal conductivity, ρ = density, and C_p = heat capacity) of materials exists, it will lead to establishment of a highly asymmetric thermal field. The thermal profiles for a given set of processing parameters are schematically shown in Figure 2.6. For similar welding, Figure 2.6A shows existence of symmetrical temperature contours on the transverse cross-section of the weld. However, an asymmetric temperature contour should be noted in Figure 2.6B due to differences in thermophysical properties of both materials. The existence of an asymmetric thermal field would cause a sharp temperature gradient as a function of distance in lower thermal diffusivity material. If it happens to be also a high-strength material with very high softening temperature, it will lead to significant differences in flow characteristics for two materials especially toward the root of the weld. It might result in a lack of bonding between two materials which will lead to defects in the root region of the weld. In such cases, it is advisable to plunge the tool biased more toward lower thermal

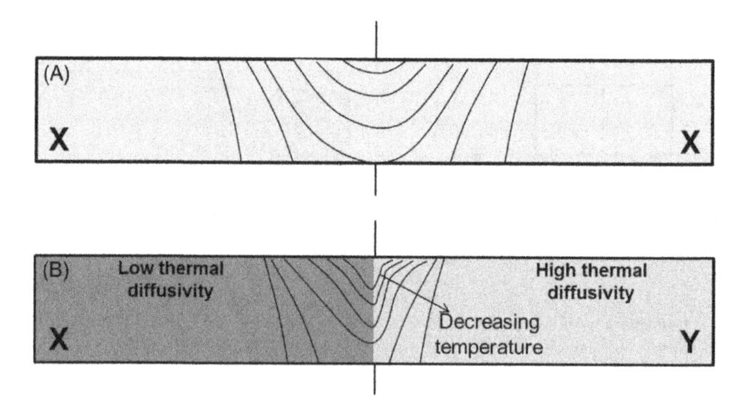

Figure 2.6 Temperature contours showing (A) symmetric profile in similar welding and (B) asymmetric profile in dissimilar welding.

diffusivity material. But, in doing so if temperature exceeds the melting point of lower melting temperature material, it will cause melting of the lower melting temperature material which will eventually lead to defective or unsuccessful weld. It is quite possible in the welding of Al/Cu alloys, but, for Fe/Cu combination, Fe being a high-temperature material, the plunging of the tool biased toward Cu alloy can be considered. Experiment coupled with thermal simulation will be useful in establishing the degree of offset of the tool to obtain a successful weld.

2.4 MATERIALS FLOW AND MIXING

The material flow model developed to understand material flow around the tool during FSW can also be utilized to understand material flow in dissimilar welding using FSW. As per wiping flow mechanism proposed by Nunes (2006), the metal from the front is wiped onto the tool and is wiped off on the trailing side of the tool. Entire episode of material transfer around the tool takes place in "first in–last out" fashion. This is shown schematically in Figure 2.7 for similar metal welding. It shows that a fresh material makes contact with the tool on the AS of the tool. As it rotates a new set of fresh materials gets deposited on the material already deposited. It results in an increase in the shear layer thickness as a particular point on the tool moves from AS to RS of the tool. This figure can be modified for dissimilar metal welding and has been shown in Figure 2.8. It shows that up to point B, only material X wipes onto the tool. At point B and onward, material Y starts wiping onto the material X, which has

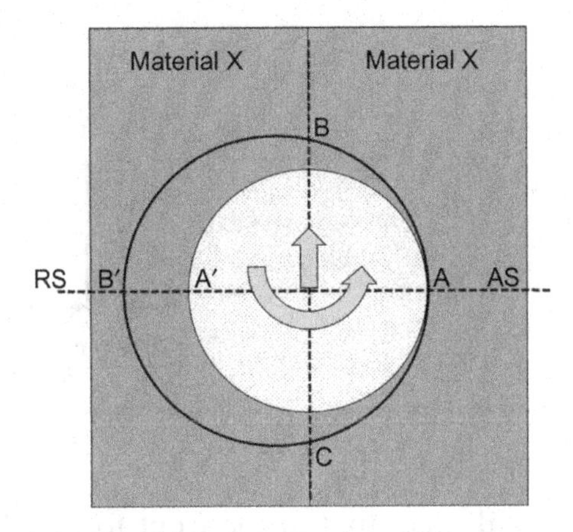

Figure 2.7 Visualization of material flow during FSW process (AS, advancing side; RS, retreating side).

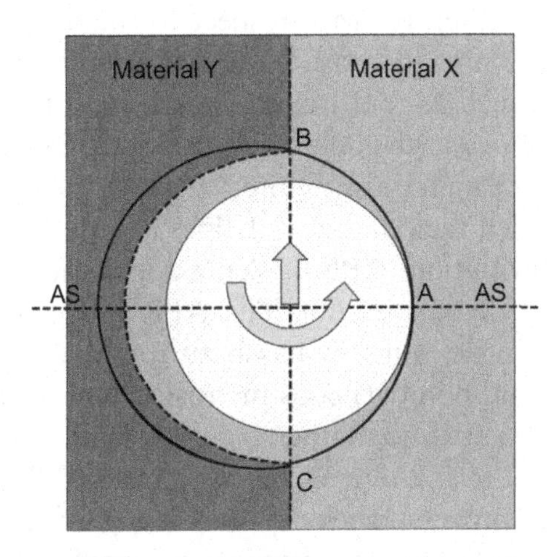

Figure 2.8 Visualization of material flow in dissimilar metal welding during FSW (AS, advancing side; RS, retreating side).

accumulated during the tool rotation between points A and B. In this simplistic description of material flow around the tool in the dissimilar metal welding, point B has been depicted to be present at weld centerline, which coincides with the linear path of the tool movement. However, in general, it does not need to be so. Point B or in other words the weld centerline can be biased toward either AS or RS. Such arrangements will have impact on material flow around the tool and the flow pattern developed eventually in the nugget of the weld.

It is also recognized based on material flow study using markers that, in general, materials present at the front of the tool toward AS

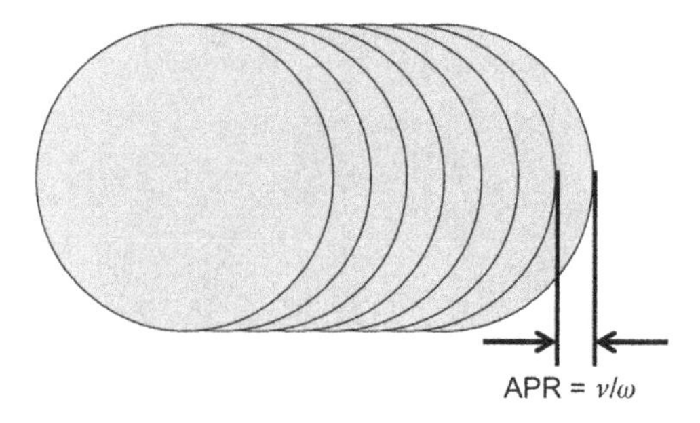

Figure 2.9 The onion ring pattern observed on the top surface of friction stir weld.

have different flow path than that at the front toward RS. Hence, the level of intermixing between materials X and Y would also depend on the position of weld interface with respect to the direction of the tool movement. In addition to the position of the weld interface, other processing parameters such as tool traverse speeds and tool rotation rates would be of importance in the determination of the level of intermixing between two different materials. Figure 2.9 shows a typical top view of the friction stir welded region, and it defines a very important parameter "advance per revolution (APR)," defined as a ratio of tool traverse speed (ν) to tool rotation rate (ω). The APR is equal to the spacing between two consecutive rings. It has been found that a low value of APR results in a more homogeneous microstructure in similar welding. A low value can be obtained either by decreasing ν or increasing ω while keeping the other parameter values constant. But, decreasing ν would mean lower process efficiency and a higher ω would mean a broader HAZ. Hence, process optimization would be needed either through design of experiments or experiments assisted with process simulation to have an acceptable level of homogenization of materials in the nugget in the course of dissimilar FSW to have a good balance of properties of the weldments.

Another important point from material mixing point of view is the extent of deformation of material around the welding tool. To illustrate this point Figure 2.10 has been included. It shows degree of deformation of materials surrounding tool by dashed-dotted line. This schematic is based on the model presented by Long et al. (2007). As per this, the material surrounding the tool in the first quadrant of the tool undergoes higher thermo-mechanical deformation compared to that in quadrant II. Although a complex state of stress exists in the

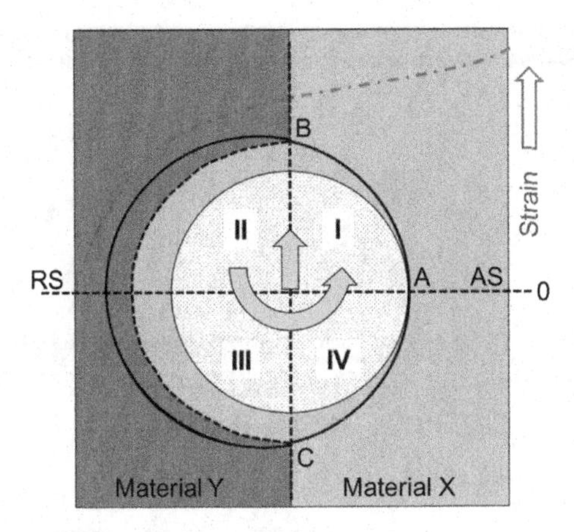

Figure 2.10 Positional dependence of degree of deformation during FSW.

material around the tool, shear is a dominant component. To effect shear-assisted materials mixing, the foregoing discussion suggests to shift the weld centerline more toward the AS of the tool. In this configuration a better mixing between materials X and Y is assisted partly due to the kinematics of the process and partly due to large shear forces existing in quadrant I of the tool.

Additional benefit of keeping the interface biased toward AS would be a higher deformation of the oxide layer on the faying surface which would cause significant refinement of the oxide film. Moreover, the presence of faying surface toward AS will result in a better distribution of these fragmented oxide particles in the nugget. In fact, even in the case of similar welding, placement of faying surface toward the AS may prove beneficial.

The experimental proof of the hypothesis being discussed here with respect to positional dependence of weld interface in relation with the tool rotation axis can be found in the work carried out by Kumar and Kailas (2010). In this work, Kumar and Kailas (2010) friction stir welded 4.4 mm thick sheets of AA7020-T6 alloy by varying the distance of weld centerline with respect to the tool rotation axis. The tool was biased toward AS of the tool from the interface during plunge stage and it was made to move at an angle to the joint line so that when the tool retracted from the welded sheet it was on RS of the weld. The results based on this study are included in Figure 2.11. It shows the transverse cross-section of the weld representing different

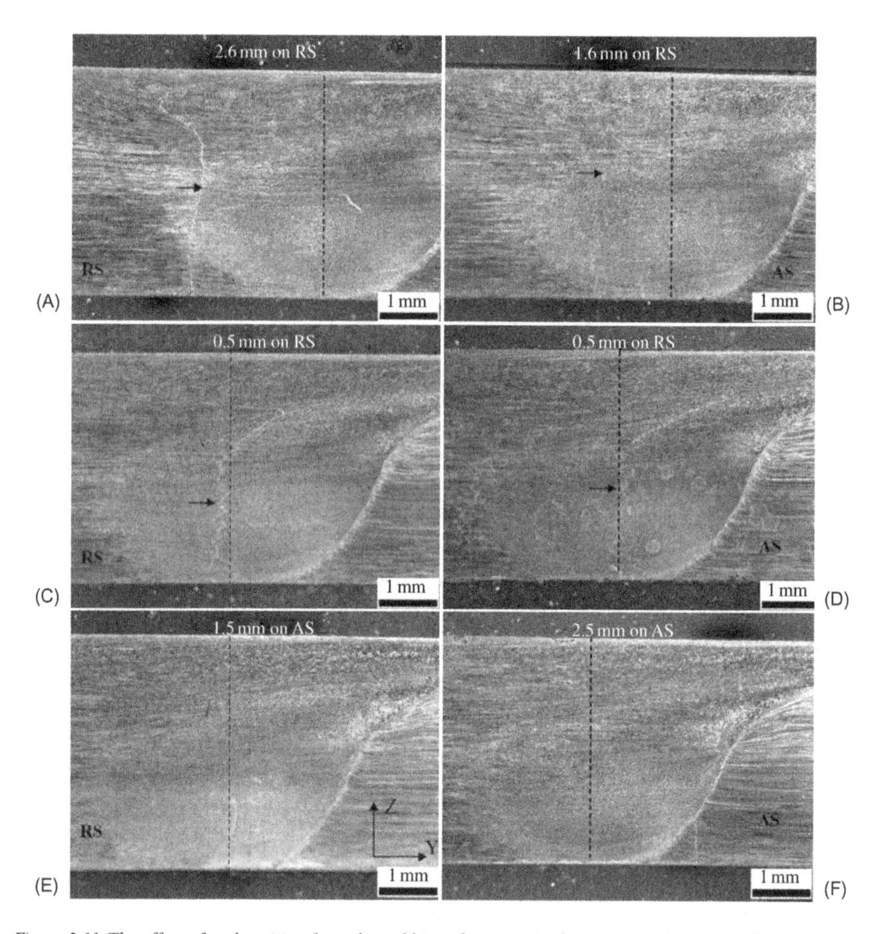

Figure 2.11 The effect of tool position from the weld interface on joint line remnant (Kumar and Kailas, 2010), Reprinted with permission from Maney Publishing.

distances of the weld interface from the tool rotation axis. It clearly shows that the tool disrupts the oxide layer on the faying surface and distributes it homogeneously once the tool center is at 2.5 mm from the weld interface on AS (Figure 2.11F).

2.5 FORMATION OF INTERMETALLIC COMPOUNDS

The formation of intermetallic phases during dissimilar welding using conventional fusion welding is an issue because it can impair the joint integrity severely depending on the thickness. In this regard, the solid-state welding has found a niche over conventional fusion welding. Although it does not eliminate the issue of intermetallic phase formation, it does certainly alleviate the issue. Figure 2.12 shows Al–Fe phase

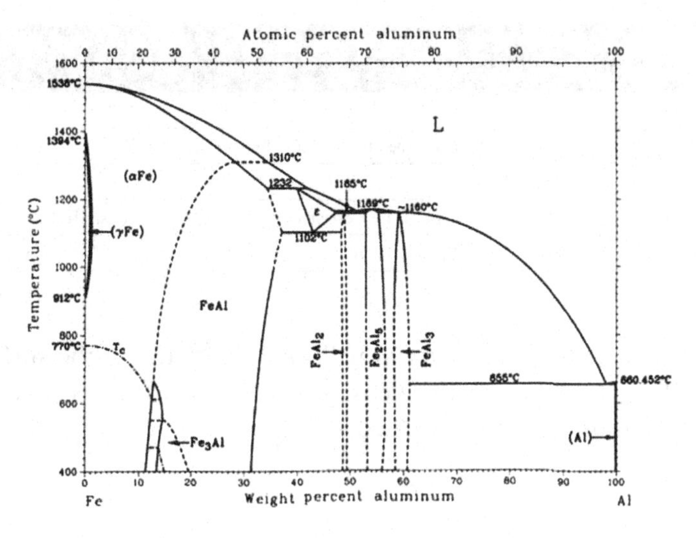

Figure 2.12 Fe–Al binary phase diagram (Kattner and Burton, 1992).

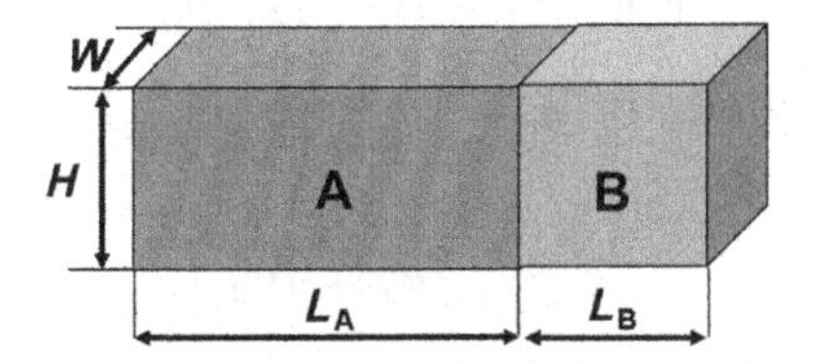

Figure 2.13 A schematic showing material volumes A and B which will mix together during FSW resulting in a global weld composition A–wt% B.

diagram. It highlights three points—(i) widely differing melting points of Al and Fe, (ii) presence of a wide variety of intermetallic phases, and (iii) negligible solubility (for Al in Fe at room temperature) of these elements in each other. We know that weld properties such as mechanical behavior is greatly affected by shape, size, type, and volume fraction of intermetallic phase(s) in the welded zone. Hence, it becomes important to develop, at least qualitatively, an understanding of factors affecting various descriptors related to intermetallic phase formation.

Figure 2.13 shows volume of materials A and B which will get mixed during FSW, and it will result in a global weld composition A–wt% B. Number and type of phases in the weld will depend on the location of this composition in binary phase diagram. A back-of-the-envelope calculation is carried out below to understand the composition of the weld during FSW. For simplicity of calculation it is assumed here that the shape of the material volume is of parallelepiped

Table 2.2 Composition of the Weld Nugget as a Result of Tool Offset with Respect to Weld Centerline During FSW

A = Al, B = Fe		
$L_{Al} + L_{Fe} = D_{pin}$ (pin diameter of the tool; assumed to be 10 mm for this calculation)		
$\rho_{Al} = 2.7$ g/cc, $\rho_{Fe} = 8.1$ g/cc		
Case I: Fe−90 wt% Al	Case II: Fe − 50 wt% Al	Case III: Fe−10 wt% Al
$L_{Fe} = 0.36$ mm	$L_{Fe} = 2.5$ mm	$L_{Fe} = 7.5$ mm

shape. In actual welding, material will take the shape of the surface of the pin. If ρ_A and ρ_B are the densities of A and B, respectively, the mass of the volume being processed,

$$W_A = W * H * L_A * \rho_A \tag{2.1}$$

and

$$W_B = W * H * L_B * \rho_B \tag{2.2}$$

Hence, weight fraction of B,

$$f_{wt,B} = \frac{W_B}{W_A + W_B} \tag{2.3}$$

Substituting Eqs. (2.1) and (2.2) into (2.3), we get

$$f_{wt,B} = \frac{1}{1 + (\rho_A/\rho_B)(L_A/L_B)} \tag{2.4}$$

Table 2.2 shows how by varying the ratio L_A and L_B one can control the composition of the welded zone.

Although very simple, the calculation carried out here indicates that when FSW tool is almost entirely plunged into aluminum, as per Figure 2.12 pure Al and intermetallic phase $FeAl_3$ will coexist in the nugget. On the extreme, when the tool is plunged mostly toward Fe such that proportion of Al is less than 10 wt%, above 400°C only single phase α-Fe(Al) exists. Hence, in dissimilar metal welding using FSW technique, tool offset becomes a necessity to control the intermetallic phases in the weld nugget. Above calculation suggests that it is possible to control the formation of intermetallic phases if offset is controlled properly. Figure 2.14 shows two different configurations of Al and Fe workpieces where a major portion of the tool pin is plunged into Fe to avoid formation of intermetallic phase(s). In case I

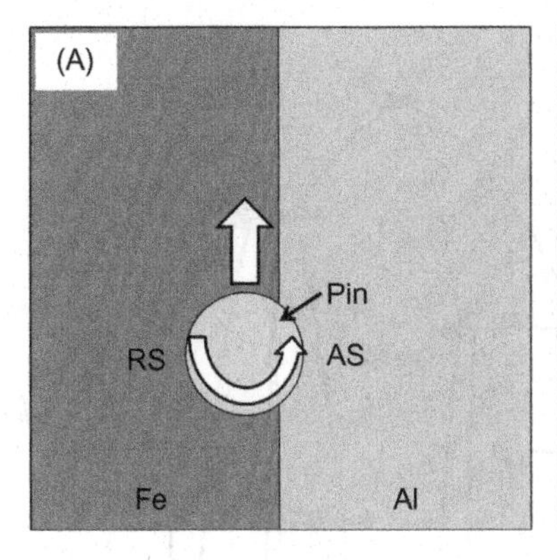

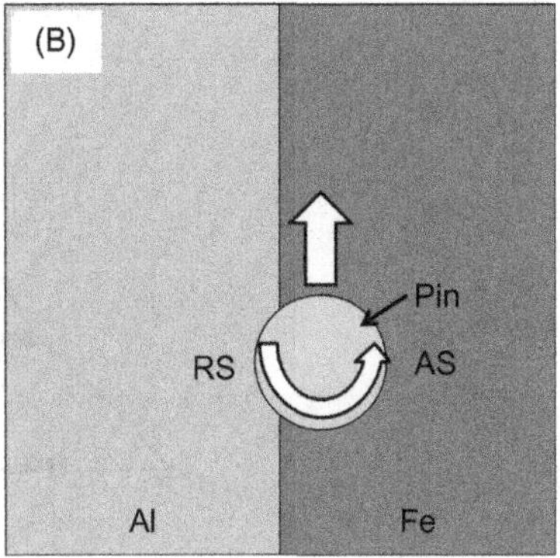

Figure 2.14 Two possible configurations of weld between Al and Fe to avoid formation of intermetallic phases.

configuration, Fe is placed on RS (Figure 2.14A) and in another on AS of the weld (Figure 2.14B). Based on the discussion regarding material and faying surface mixing, the welding configuration presented in Figure 2.14A will result in a homogeneous microstructure in the weld nugget without any intermetallics. But, in reality this configuration has hardly been tried. The configuration which has been extensively tried in making welds between Al alloys and steels correspond to case II but a major part of the tool pin plunged into Al side of the workpiece. From Fe–Al binary phase diagram, this corresponds to Al-rich end of the phase diagram which would eventually lead to formation of intermetallic phase. The reason behind not trying the configuration corresponding to case I (Figure 2.14A) is generation of intense heat during welding when tool plunges mostly into Fe side. The heat causes temperature to rise above the melting point of the aluminum, and it results in melting of aluminum abutting the Fe workpiece. It leads to a defective or no weld formation between Al and Fe.

2.5.1 Mechanism of Intermetallic Phase Formation

The formation of intermetallic phases during dissimilar metal welding is somewhat similar to intermetallic phase formation during surface composite formation using friction stir processing technique. Hence, here Al–Ni system will be considered for discussion of the formation of intermetallic phase during FSW. An Al–Ni binary phase diagram is included in Figure 2.15. It shows that there are five intermetallic phases and two terminal solid solutions below 640°C. The formation, type, and number of phases would depend on the weight fractions of Al and Ni. For example, if Al and Ni are mixed in such a way that the

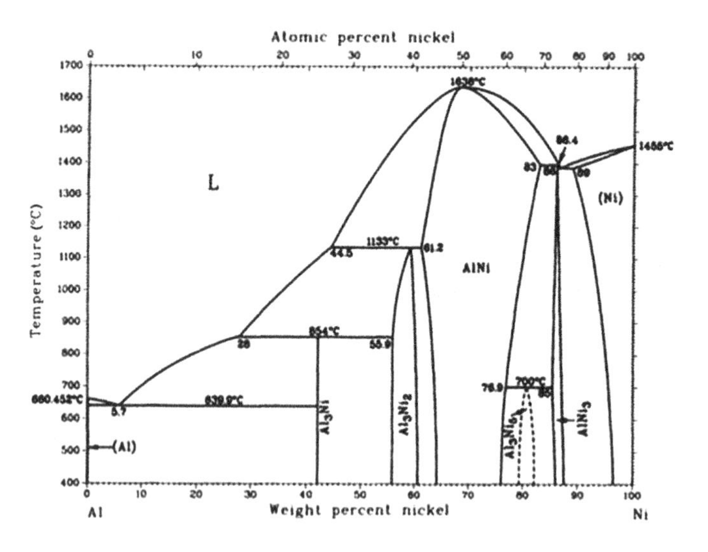

Figure 2.15 Binary Al–Ni phase diagram (Nash et al. (1992)).

overall composition is below 40–42 wt% Ni, the alloy in equilibrium condition would consist of Al and Al_3Ni intermetallic phase. However, in diffusion couple experiments and in a solid-state process like FSW/P, the formation of different phases is a nonequilibrium process. Qian et al. (2012) have discussed the formation of intermetallic phases in aluminum alloy reinforcement by *in-situ* formed Al_3Ni intermetallic during friction stir processing. Qian et al. (2012) considered effective Gibbs free energy change of formation (ΔG_p^e) for predicting the formation of the type of intermetallic phase. For formation of an intermetallic phase p,

$$\Delta G_p^e = \Delta G_p \times \frac{C_e}{C_p} \tag{2.5}$$

where ΔG_p, C_e, and C_p are Gibbs free energy of change of formation for the phase p, the effective concentration of the limiting element at the interface, and the concentration of the limiting element in the intermetallic compound, respectively. Table 2.3 provides the thermodynamic data for the calculation of ΔG_p^e for two compositions selected at random. For $Al_{0.5}Ni_{0.5}$, the most negative value is found for AlNi intermetallic phase. Hence, in this condition the formation of this phase will be favored at 430°C (the peak temperature reported by Qian et al., 2012). For $Al_{0.965}Ni_{0.035}$, the phase $NiAl_3$ showed the most negative value of ΔG_p^e. It, therefore, will be the phase formed at the

Table 2.3 Calculation of Effective Gibbs Free Energy for Two Different Compositions of Al–Ni Binary System at 703 K

Effective Concentration			$Al_{0.50}Ni_{0.50}$		$Al_{0.965}Ni_{0.035}$	
Phase	ΔG_i (*T*) (J/mol)	ΔG_i (703 K) (J/mol)	Limiting Elements	ΔG_i^e (kJ/mol)	Limiting Elements	ΔG_i^e (kJ/mol)
$Ni_3Al(Ni_{0.75}Al_{0.25})$	$-40246.5 + 6.24T$	−35.86	Ni	−23.91	Ni	−1.67
$NiAl(Ni_{0.50}Al_{0.50})$	$-76198.7 + 13.2T$	−66.92	Ni/Al	−66.92	Ni	−4.68
$Ni_2Al_3(Ni_{0.40}Al_{0.60})$	$-71545.2 + 13.7T$	−61.94	Al	−51.6	Ni	−5.42
$NiAl_3(Ni_{0.25}Al_{0.75})$	$-48483.8 + 12.3T$	−39.84	Al	−26.56	Ni	−5.58
Source: *Reproduced from Qian et al. (2012).*						

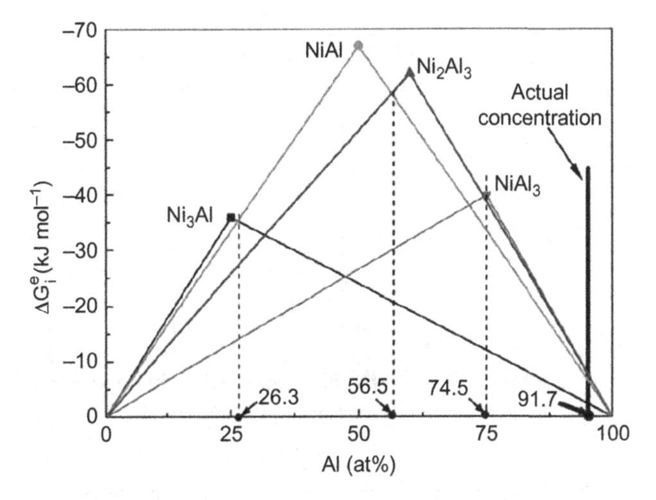

Figure 2.16 The variation of effective Gibbs free energy of various intermetallic phases found in Al–Ni binary system (Qian et al., 2012).

interface of Al and Ni phases in this condition. To assess the variation of intermetallic phase formation as a function of effective composition, Eq. (2.5) has been represented graphically in Figure 2.16 for different intermetallic phases found in Al–Ni binary system. Figure 2.16 shows that when concentration of Al is less than 26.3 at%, it favors formation of Ni_3Al phase. Similarly, when compositions are in the range 26.3–56.5, 56.5–74.5, and above 74.5, formation of NiAl, Ni_2Al_3, and $NiAl_3$, respectively, should be expected. Hence, this simple approach can be taken to evaluate the formation of first intermetallic phase during dissimilar metal welding. Clearly, the offset method used in FSW of dissimilar materials will cause effective concentration of the nugget to change, and it will have bearing on the type of intermetallic formed during the welding. In addition to this, number of passes and post-weld heat treatment may influence the volume fraction and type of intermetallic formed. For example, only Al_3Ni phase was formed during the friction stir processing of Al–Ni system by Ke et al. (2010). When friction stir processed region was subjected to further heat treatment at 550°C for 6 h, X-ray diffraction (XRD) detected the presence of Al_3Ni_2 phase in addition to Al_3Ni phase. Figure 2.17 shows the XRD results based on the friction stir processing study carried out by Ke et al. (2010) on Al–Ni system.

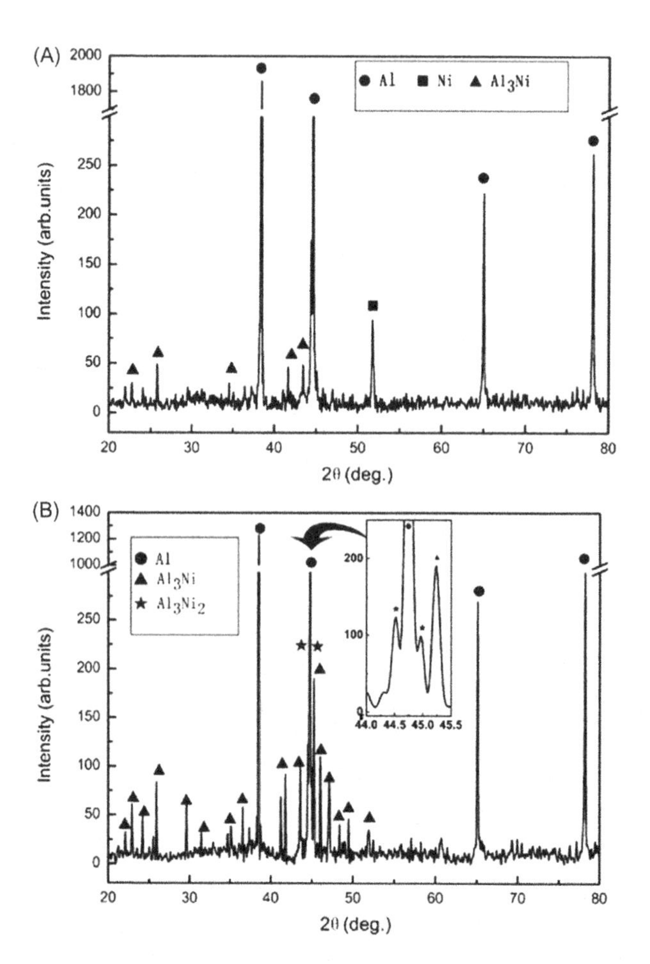

Figure 2.17 X-ray diffraction pattern for (A) three pass friction stir processed Al–Ni composite and (B) showing effect of heat treatment (at 550°C, 6 h) on the formation of intermetallic phases (Ke et al., 2010).

REFERENCES

Ashby, M.F., 2005. Materials Selection in Mechanical Design, third ed. Butterworth-Heinemann, Elsevier, Burlington MA USA.

Frick, J.P., 2000. Woldman's Engineering Alloys, ninth ed. ASM International, Materials Park, OH.

Kattner, U.R., Burton, B.P., 1992. Al (Aluminum) Binary Alloy Phase Diagrams, Alloy Phase Diagrams, vol. 3. ASM Handbook, ASM International, 2.4–2.56.

Ke, L., Huang, C., Xing, L., Huang, K., 2010. Al–Ni intermetallic composites produced in situ by friction stir processing. J. Alloys Comp. 503 (2), 494–499.

Kumar, K., Kailas, S.V., 2010. Positional dependence of material flow in friction stir welding: analysis of joint line remnant and its relevance to dissimilar metal welding. Sci. Technol. Weld. Join. 15 (4), 305–311.

Long, T., Tang, W., Reynolds, A.P., 2007. Process response parameter relationships in aluminium alloy friction stir welds. Sci. Technol. Weld. Join. 12 (4), 311–317.

Mills, K.C., 2002. Recommended Values of Thermophysical Properties for Selected Commercial Alloys. Woodhead Publishing.

Mishra, R.S., 2008. Preface to the viewpoint set on friction stir processing. Scr. Mater 58 (5), 325–326.

Mishra, R.S., De, P.S., Kumar, N., 2014. Friction Stir Welding and Processing: Science and Engineering. Springer.

Nash, P., Singleton, M.F., Murray, J.L., 1992. Al (Aluminum) Binary Alloy Phase Diagrams, Alloy Phase Diagrams, vol. 3. ASM Handbook, ASM International, pp. 2.4–2.56.

Nunes, A.C., Jr. 2006. Metal flow in friction stir welding, The Minerals, Metals, and Materials Society, 15–19 Oct. 2006 Cincinnati OH USA. Available at ntrs.nasa.gov.

Qian, J., Li, J., Xiong, J., Zhang, F., Lin, X., 2012. *In situ* synthesizing Al_3Ni for fabrication of intermetallic-reinforced aluminum alloy composites by friction stir processing. Mater. Sci. Eng. A 550, 279–285.

CHAPTER 3

Tool Design for Friction Stir Welding of Dissimilar Alloys and Materials

The essentials of friction stir welding (FSW) are plunging and stirring a hard tool generally consisting of a shoulder and a pin or probe into workpieces to be welded. The workpieces are joined together through heating, material movement, and forging dominated by the tool geometry in addition to the welding parameters. The welding tool has three primary functions: (i) heating of the workpiece, (ii) movement of material to produce the joint, and (iii) containment of the hot metal beneath the tool shoulder. Heating is created both by friction between the rotating tool (pin at initial plunge stage and mainly shoulder during the run) and the workpiece, and by severe plastic deformation of the workpiece. The localized heating softens material around the pin and, combined with the tool rotation and translation, leads to movement of material from the front to the back of the tool, thus filling the hole in the tool wake as the tool moves forward. The tool shoulder restricts metal from flowing out and applies forging pressure to consolidate the material right behind the moving pin (Mishra and Mahoney, 2007).

Tool design is the most influential aspect of FSW process development. Figure 3.1 is a flow chart of general approach for process development. Removal of any welding defects, such as voids, incomplete root penetration, is the prerequisite for reliable welds. For a given joint design (type of workpiece material, thickness, and joint configuration), proper tool material and tool geometry can be specified, even though the understanding has been mainly empirical in nature. For similar material FSW, solid metallurgical bonding between workpieces is easily achievable. This becomes challenging when dissimilar materials with different chemical and physical properties, especially quite different melting temperatures and thermal conductivities, are subjected to FSW. Placing of dissimilar workpieces on advancing or retracting side (in butt welding), in upper or lower sheet (in lap welding) and plunging

Friction Stir Welding of Dissimilar Alloys and Materials. DOI: http://dx.doi.org/10.1016/B978-0-12-802418-8.00003-5

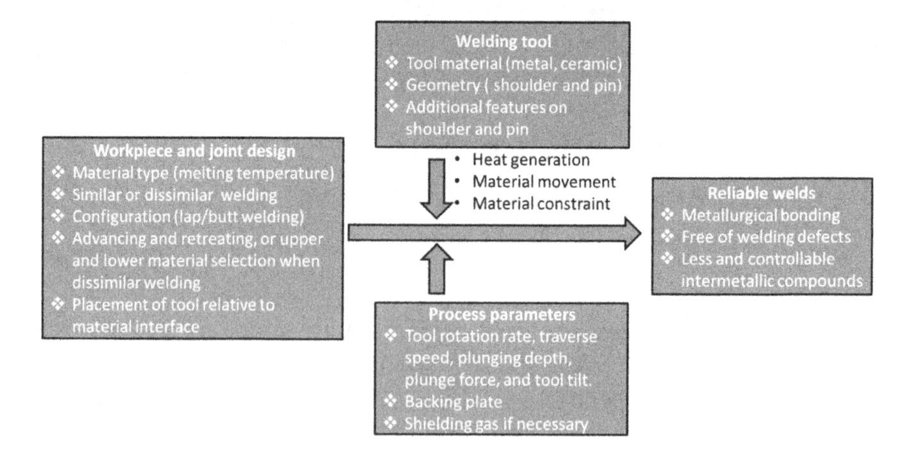

Figure 3.1 Flow chart of general approach for FSW process development.

rotating tool to a specific location relative to the joint interface, alters the weldability and impacts tool life. For dissimilar materials with high chemical affinity at elevated temperatures, which give rise to the easy formation of low-melting-temperature eutectic structure and rapid growth of brittle intermetallic compounds (IMCs), an effective tool design could play an important role in preventing the liquid film formation and reducing the formation of IMC.

There have been a number of publications on FSW of dissimilar materials in the last decade, which have mainly focused on the welding of Al alloys to low- and high-melting-temperature materials for combined structural properties. There are also some reports dealing with dissimilar FSW of high-melting-temperature materials including steels, and titanium alloys. The following sections mainly consider tooling materials and tool geometry design for dissimilar FSW.

3.1 TOOL MATERIALS COMPARED TO WORKPIECES

Weld quality, tool wear, and cost are important considerations in the selection of tool material. In other words, the selection mainly depends on the workpiece material to be welded and cost-effectiveness of tool material and associated machining cost to desired shape. How easily the workpieces can be welded without defects and detrimental IMC with acceptable tool life specifies the requirement on tool material.

Tool steel is commonly selected as tooling material when dissimilar FSW is performed for low-melting-temperature materials, such as Al and Mg alloys. Open literature on welding of dissimilar Al alloys and Al to Mg alloys shows a significant reliance on tool steels as tool material. The advantages of the use of tool steels for dissimilar welding of relatively low-melting-temperature materials include easy availability and machinability, low cost, established material characteristics, and mild or negligible tool wear. Tool steel, especially H13, a chromium–molybdenum hot-worked air-hardening steel has been most frequently adopted due to its good elevated-temperature strength, thermal fatigue resistance, and wear resistance. Tool steels such as SKD51, O1, and high-speed steel have also been reported frequently for dissimilar low-melting-temperature material FSW.

Wear of tool needs to be considered when welding is performed between low- and high-melting-temperature materials, such as Al or Mg alloys to steels or titanium alloys. Particularly in this case, the welding process design is critical; whether tool contacts the high-melting-temperature material and how much it penetrates into the high-melting-temperature material influence the tool wear. In general, active material mixing between dissimilar workpieces via pin stirring and scribing is required for mechanical performance by preventing IMC thickening and forming either a metallurgical bonding or a mechanical interlocking. Depending on the grade of the high-melting-temperature materials to be welded, tool steels (Elrefaey et al., 2005; Watanabe et al., 2006; Chen et al., 2008) and alloy steels (Chen, 2009) have been used in both lap and butt configurations. Chen et al. (2013) lap-welded AA6060 to mild steel using tool steel as tool material and reported no significant wear of tool pin even when the pin was plunged into the mild steel by 0.1 mm. Haghshenas et al. (2014) welded AA5754 Al alloy with high-strength steels DP600 and 22MnB5 in lap joint configuration using tool steel as tool material without its excessive wear by placing the softer Al alloy on top of the steel plates and avoiding direct contact of the tool with the steel plates. In butt joint configuration, the higher melting temperature workpiece is often placed on the advancing side and the welding tool is offset from the butt interface toward the lower melting temperature material to prevent tool wear and overheating of the lower melting temperature materials.

Tungsten-based tools such as tungsten rhenium and tungsten carbide are necessary, at least for the pin portion if the tool will be plunged in the hard material and subjected to severe frictional conditions. Liyanage et al. (2009) used W−25Re tool to make dissimilar friction stir spot welds between Al alloy and steel, and between Mg alloy and steel with report of some tool wear. Bozzi et al. (2010) applied W−25Re to friction stir spot welded AA6016 and IF steel. Xiong et al. (2012) friction stir lap-welded AA1100 and SUS321 stainless steel using a cutting pin made of WC-Co which exhibited good wear resistance. Liu et al. (2014) butt-welded AA6061 and high-strength TRIP 780/800 steel using WC-10%Co by slightly offsetting pin to Al side. High-temperature-resistant coatings can further prevent tool wear. Da Silva et al. (2010) friction stir spot welded AA1050 and hot stamped boron steel using a WC-Co tool with a 3 μm thick AlCrN coating and reported negligible wear after 32 spots.

In addition to tungsten-based material, a polycrystalline cubic boron nitride (PCBN) is a preferred tool material for deep insertion welding of high-temperature materials, such as steels, Ti alloys, and nickel-base superalloys. Due to its high strength and hardness at elevated temperatures along with high-temperature stability, the PCBN pin could resist wear when it was plunged into the lower steel deeply. Sawada and Nakamura (2009) successfully lap-welded 3 mm thick stainless steel 304 to a same thickness ductile cast iron FCD450 using a PCBN tool by preheating of the workpieces. Choi et al. (2010) butt-welded a 4 mm thick low-carbon steel SPHC to a high-carbon steel SK85 with PCBN tool and reported superior tensile property comparable to SPHC. Although the performance of PCBN tool is promising for welding high-temperature materials, the cost for making PCBN tool is very high due to critical manufacturing processes which require high temperatures and pressures.

Figure 3.2 is a summary of the current tool materials used for friction stir welding of the indicated dissimilar materials.

3.2 INFLUENCE OF TOOL GEOMETRY ON MATERIAL FLOW CONTROL

Tool geometry affects the frictional heat generation and plastic material flow, in turn, joint integrity, resulting microstructure and mechanical properties. Compared to similar FSW, the tool geometry plays a more critical role in dissimilar materials FSW. When designing a tool for dissimilar FSW, attention needs to be paid to defect formation and

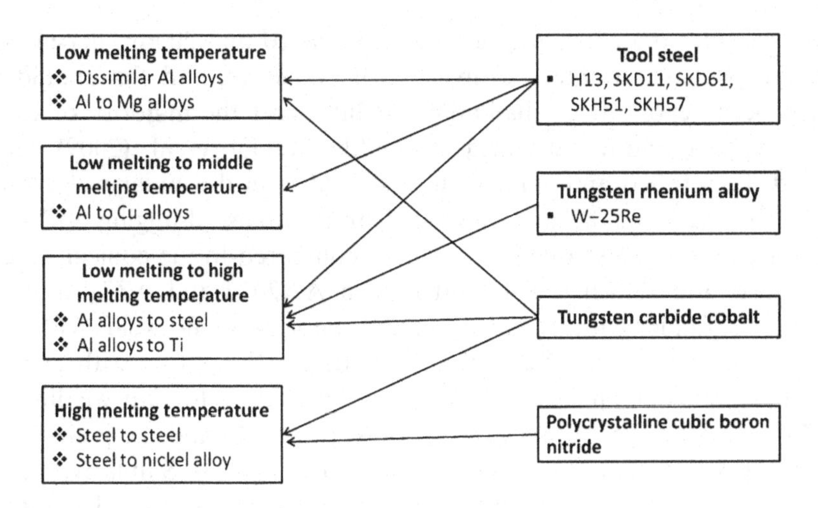

Figure 3.2 Summary of the most frequently used tool materials for dissimilar FSW.

IMC formation. Due to the asymmetrical temperature generation during friction stir and intrinsic physical property differences, it is expected that the material flow in the weld zone would be more inhomogeneous compared to similar FSW and more challenging to achieve defect-free welds. For dissimilar materials with high chemical affinity at elevated temperatures, formation of IMC is generally not avoidable. Conceptually a reaction layer of finite thickness will form. However, if the type, thickness, or morphology of IMC can be optimized, it will have beneficial impact on mechanical performance.

Both defect and IMC formations are directly linked to material mixing. Is material mixing desirable for sound dissimilar material welds? In general, active material mixing between dissimilar workpieces via pin stirring and scribing is required for mechanical performance by disrupting the tenacious oxide layer present on the surfaces, preventing IMC thickening and forming either a metallurgical bonding or a mechanical interlocking. It may be desirable to have a different degree of mixing for dissimilar material systems as compared to that utilized in FSW of similar materials. The design of the tool influences the degree to which the two materials are mixed during welding. Important factors include shoulder diameter, shoulder surface feature, pin shape, size, and additional surface features.

Tool shoulders are designed to produce heat through friction and material deformation to the surface and subsurface of the workpiece.

The shoulder diameter is generally between 10 and 20 mm depending on the pin size adopted. Convex scroll, concave, and flat shoulders have been reported for dissimilar welding, and the majority of tools used have a traditional concave shoulder. Leitão et al. (2009) compared the scrolled and concave tool shoulder on dissimilar welding of AA6016 and AA5182 and suggested more intense dragging and stirring action for the scrolled shoulder as compared to the conical shoulder. Cao and Jahazi (2010) butt-welded AA2024 and AZ31 using a scrolled shoulder with threaded pin and reported sound welds with low distortion and no solidification cavities or cracks. A tool with convex scrolled shoulder and stepped spiral pin was used for lap welding of AZ31 Mg alloy and a high-strength low alloy steel and the joints were found to be mechanical in nature (Jana et al., 2010). Similarly, AZ31 and zinc-coated mild steel were lap joined by using a scrolled shoulder with a square shape pin (Schneider et al., 2011). Tool features can influence the IMC formation. Galvão et al. (2010) butt-welded AA5083 and Cu-DHP by using scrolled and conical shoulder tools and found that tool shoulder deeply influences the distribution of Al and Cu in the weld and the formation of IMCs. The scrolled shoulder tool results in through thickness material mixing and formation of a mixed region almost exclusively composed of $CuAl_2$. The conical tool, on the other hand, introduced highly heterogeneous mixture of Al, Cu, $CuAl_2$, and Cu_9Al_4 in the nugget with lower intermetallic content.

Pin features are much more complex compared to those of the shoulder. The tool pin is designed to disrupt the faying surface of the workpiece. Features such as threads, steps, flats, or flutes have been widely used for various alloy systems to control the material flow for better material mixing. It has been shown that flats on the pin surface impose vertical material flow which can promote intermixing. Zettler et al. (2006) have shown that by employing conical pin tools with flats, a more flexible placement of the workpieces can be made with good material mixing during dissimilar welding of AA2024 and AA7050. Pins with cutting feature have been developed for lap welding of low- and high-melting-temperature material by scribing the high-melting-temperature material to create either metallurgical bonding or mechanical interlocking. Jana and Hovanski (2012) lap-welded AZ31 Mg alloy and a high-strength low alloy steel using a stepped spiral pin with a short WC insert at the pin bottom. The short WC insert was plunged into the lower steel and that acted as a cutter to deform the steel and mix it with upper Mg. Xiong et al. (2012) reported sound lap welds between AA1060 and stainless steel SUS321 using a tungsten carbide pin with cutting edges. Same pin design has been applied to dissimilar lap welding of AA1060 to titanium alloy Ti-6Al-4V (Wei et al., 2012a) and AZ31 to stainless steel SUS321 (Wei et al., 2012b).

Another important factor for dissimilar FSW is the pin length. For butt welding, the pin length is generally very close to the total thickness of workpieces. In terms of lap welding, a pin length close to or slightly longer than the thickness of upper sheet workpiece is recommended for disrupting the surface oxide, creating intimate contact between fresh metal surfaces, and limiting IMC formation, in turn, a sound weld with good mechanical performance.

In summary, when it comes to tool design for FSW of dissimilar materials, a tool material can be selected based on the design need. Regular tool steel can be chosen as tool material if the tool mainly plunges into lower melting materials even for a combination between low- and high-melting-temperature materials. Tungsten-based tools or even expensive PCBN tools may be employed if severe tool wear is expected, especially when welding is conducted between dissimilar high-temperature materials. In terms of tool geometry design, it mainly depends on the needs for material mixing. Shoulder features such as convex scrolls and concave are general options. Pin features such as flats, threads, and steps are also believed to enhance material mixing. Additional cutter or insert may be added to tool pin to improve mixing.

REFERENCES

Bozzi, S., Helbert-Etter, A.L., Baudin, T., Criqui, B., Kerbiguet, J.G., 2010. Intermetallic compounds in Al 6016/IF-steel friction stir spot welds. Mater. Sci. Eng. A 527, 4505–4509.

Cao, X., Jahazi, M., 2010. Friction stir welding of dissimilar AA 2024-T3 to AZ31B-H24 alloys. Mater. Sci. Forum 638-642, 3661–3666.

Chen, T., 2009. Process parameters study on FSW joint of dissimilar metals for aluminum-steel. J. Mater. Sci. 44, 2573–2580.

Chen, Y.C., Komazaki, T., Kim, Y.G., Tsumura, T., Nakata, K., 2008. Interface microstructure study of friction stir lap joint of AC4C cast aluminum alloy and zinc-coated steel. Mater. Chem. Phys. 111, 375–380.

Chen, Z.W., Yazdanian, S., Littlefair, G., 2013. Effects of tool positioning on joint interface microstructure and fracture strength of friction stir lap Al-to-steel welds. J. Mater. Sci. 48, 2624–2634.

Choi, D.H., Lee, C.Y., Ahn, B.W., Yeon, Y.M., Park, S.H.C., Sato, Y.S., et al., 2010. Effect of fixed location variation in friction stir welding of steels with different carbon contents. Sci. Technol. Weld. Join. 15, 299–304.

Da Silva, A.A.M., Aldanondo, E., Alvarez, P., Arruti, E., Echeverría, A., 2010. Friction stir spot welding of AA 1050 Al alloy and hot stamped boron steel (22MnB5). Sci. Technol. Weld. Join. 15, 682–687.

Elrefaey, A., Gouda, M., Takahashi, M., Ikeuchi, K., 2005. Characterization of aluminum/steel lap joint by friction stir welding. J. Mater. Eng. Perform. 14, 10–17.

Galvão, I., Leal, R.M., Loureiro, A., Rodrigues, D.M., 2010. Material flow in heterogeneous friction stir welding of aluminium and copper thin sheets. Sci. Technol. Weld. Join. 15, 654–660.

Haghshenas, M., Abdel-Gwad, A., Omran, A.M., Gökçe, B., Sahraeinejad, S., Gerlich, A.P., 2014. Friction stir weld assisted diffusion bonding of 5754 aluminum alloy to coated high strength steels. Mater. Des. 55, 442–449.

Jana, S., Hovanski, Y., 2012. Fatigue behaviour of magnesium to steel dissimilar friction stir lap joints. Sci. Technol. Weld. Join. 17, 141–145.

Jana, S., Hovanski, Y., Grant, G.J., 2010. Friction stir lap welding of magnesium alloy to steel: a preliminary investigation. Metall. Mater. Trans. A 41, 3173–3182.

Leitão, C., Emílio, B., Chaparro, B.M., Rodrigues, D.M., 2009. Formability of similar and dissimilar friction stir welded AA 5182-H111 and AA 6016-T4 tailored blanks. Mater. Des. 30, 3235–3242.

Liu, X., Lan, S., Ni, J., 2014. Analysis of process parameters effects on friction stir welding of dissimilar aluminum alloy to advanced high strength steel. Mater. Des. 59, 50–62.

Liyanage, T., Kilbourne, J., Gerlich, A.P., North, T.H., 2009. Joint formation in dissimilar Al alloy/steel and Mg alloy/steel friction stir spot welds. Sci. Technol. Weld. Join. 14, 500–508.

Mishra, R.S., Mahoney, M.W. (Eds.), 2007. Introduction, Friction Stir Welding and Processing. ASM International, Materials Park, OH (Chapter 1).

Sawada, Y.K., Nakamura, M., 2009. Lapped friction stir welding between ductile cast irons and stainless steels. J. Jpn. Weld. Soc. 27, 176–182.

Schneider, C., Weinberger, T., Inoue, J., Koseki, T., Enzinger, N., 2011. Characterisation of interface of steel/magnesium FSW. Sci. Technol. Weld. Join. 16, 100–106.

Watanabe, T., Takayama, H., Yanagisawa, A., 2006. Joining of aluminum alloy to steel by friction stir welding. J. Mater. Process. Technol. 178, 342–349.

Wei, Y., Li, J., Xiong, J., Huang, F., Zhang, F., Raza, S.H., 2012a. Joining aluminum to titanium alloy by friction stir lap welding with cutting pin. Mater. Charact. 71, 1–5.

Wei, Y., Li, J., Xiong, J., Huang, F., Zhang, F., 2012b. Microstructures and mechanical properties of magnesium alloy and stainless steel weld-joint made by friction stir lap welding. Mater. Des. 33, 111–114.

Xiong, J.T., Li, J.L., Qian, J.W., Zhang, F.S., Huang, W.D., 2012. High strength lap joint of aluminium and stainless steels fabricated by friction stir welding with cutting pin. Sci. Technol. Weld. Join. 17, 196–201.

Zettler, R., Da Silva, A.A.M., Rodrigues, S., Blanco, A., Dos Santos, J.F., 2006. Dissimilar Al to Mg alloy friction stir welds. Adv. Eng. Mater. 8, 415–421.

CHAPTER 4

Friction Stir Welding of Dissimilar Alloys

The advent of friction stir welding (FSW) not only brought a radical change in the joining of some similar non-weldable alloys like 2XXX and 7XXX series aluminum alloys but also enabled dissimilar welding of various disparate combinations of metals and alloys. The primary reason for the success of FSW has been the ease with which it can be used (refer to Table 1.1 for assessing the benefit of FSW). With respect to dissimilar metal welding two factors have contributed significantly—(i) lower temperature (below solidus temperature) attained during FSW and (ii) intense shear forces acting on the material surrounding the tool used for FSW. Nonetheless, FSW presents its own set of challenges when it comes to dissimilar metal welding. The advantages and disadvantages of dissimilar FSW will be discussed in the following sections where various combinations of dissimilar metal welds are discussed.

In Section 2.1, we classified all combinations of dissimilar metal welds into three broad categories: (I) same base metals but different chemistries, (II) different base metals but somewhat similar melting points, and (III) different base metals and large difference in melting points. For the presentation of results, category I has been termed as "dissimilar alloys" and results related to this are presented here. Combined categories II and III have been termed as "dissimilar materials" and results related to them have been presented in Chapter 5.

4.1 DISSIMILAR ALLOYS

4.1.1 Aluminum Alloys

Figure 4.1A shows friction stir dissimilar welds between A319 (Al–Si–Cu) and A356 (Al–Si–Mg) alloys. The weld shown here corresponds to 1120 rpm and 80 mm/min. Published results on dissimilar FSW indicate that the position of the workpiece has significant

Friction Stir Welding of Dissimilar Alloys and Materials. DOI: http://dx.doi.org/10.1016/B978-0-12-802418-8.00004-7

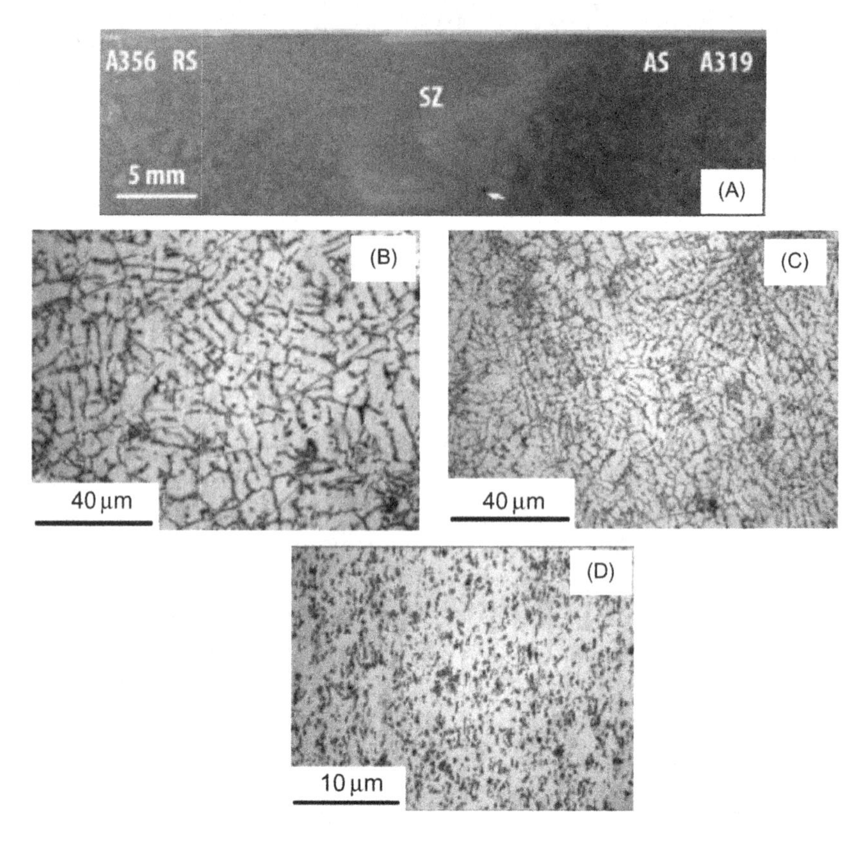

Figure 4.1 (A) Friction stir-welded A319/A356 weld, optical micrographs showing dendritic microstructure for (B) as-cast A319 (Al–Si–Cu), (C) as-cast A356 (Al–Si–Mg), and (D) distribution of Si particles in the nugget of dissimilar A319/A356 friction stir weld (Hassan et al., 2010, Reprinted with permission from Maney Publishing).

influence on material flow and joint properties of the weld. This aspect will be discussed later in greater detail. Here, in this case, only one configuration of workpieces was tried. A356 and A319 were placed on retreating side and advancing side, respectively. It shows a small cavity on the advancing side, toward the root of the weld. It is probably a lack of fill defect due to insufficient material consolidation in that region. Figure 4.1B and C shows optical micrographs of A319 and A356 alloys in as-cast condition. Clearly, both have dendritic microstructure where dendritic arm spacing in A356 is much finer than that in A319. The micrograph shown in Figure 4.1D corresponds to as-welded condition. The as-welded micrograph shows a distinctly different microstructure in the nugget. The dendritic microstructure was eliminated completely, and long acicular particles were replaced with very fine fragmented Si particles.

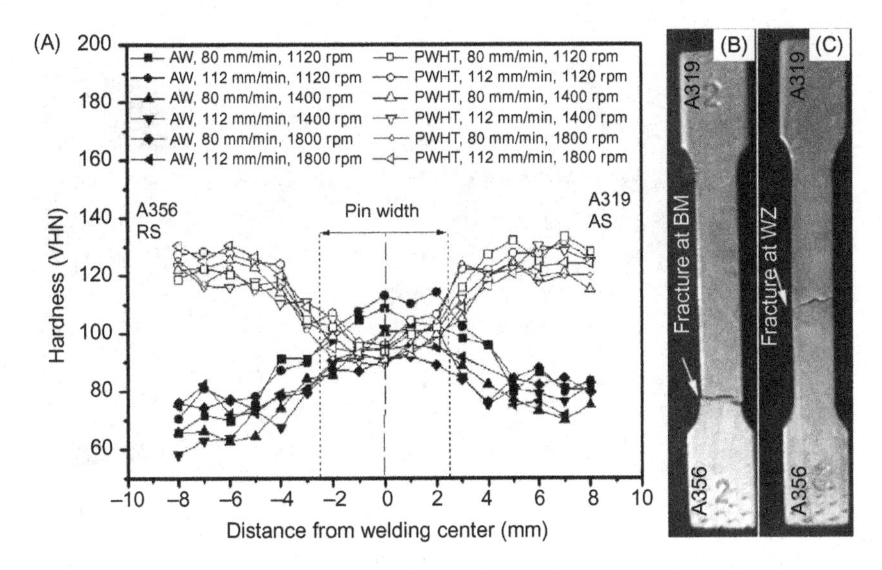

Figure 4.2 (A) Microhardness measurement profiles on the transverse cross-section of the dissimilar friction stir weld A319/A356 for different combinations of tool rotation rates and tool transverse speed. Fractured tensile samples for the weld made at 1400 rpm and 80 mm/min, (B) as-welded specimen and (C) post-weld heat treated specimen (solutionizing temperature 540°C for 12 h followed by aging at 155°C for 6 h) (Hassan et al., 2010, Reprinted with permission from Maney Publishing).

The microhardness profiles across the weld corresponding to different combinations of tool rotation rates and tool traverse speeds are shown in Figure 4.2A. The welded region, for each set of processing conditions, shows maximum hardness in the nugget, and lowest hardness values are found in the base materials. The appearance of the hardness profile can be considered symmetric across the weld centerline. On both sides of the weld centerline hardness reduction is gradual. After post-weld heat treatment, the hardness in the base materials increased; but, there was almost no change in the hardness values at the center of the nugget. However, the heat-affected zone (HAZ) shows a gradual increase in hardness from the nugget to the base materials. Since base materials were weaker than the welded zone in as-welded samples processed using 1400 rpm and 80 mm/min, the fracture took place in the base material during uniaxial tensile testing. The location of the fracture in base material was in A356 since it has lower strength than A319 (Figure 4.2B). After post-weld heat treatment, the location of the fracture moved from base material to the weld nugget of the specimen as it became the weakest zone in the welded structure (Figure 4.2C).

Figure 4.3 includes the dissimilar weld made between cast A356 and wrought AA6061. Figure 4.3A and B describes two different situations.

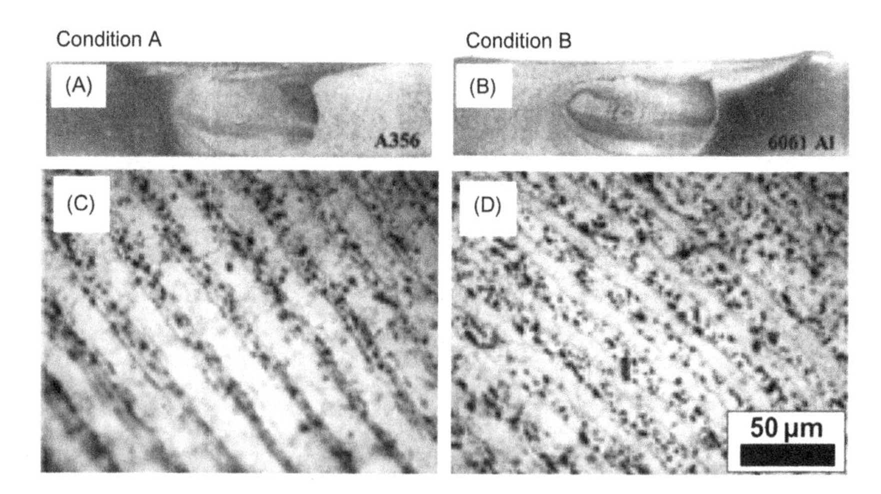

Figure 4.3 Transverse cross-section of the weld between cast A356 and wrought AA6061 fabricated at 1600 rpm tool rotation rate and 87 mm/min tool traverse speed; (A) A356 on advancing side, (B) AA6061 on advancing side, (C) an optical micrograph for weld shown in (A), and (D) an optical micrograph for the weld shown in (B). (C and D) show the presence of A356 (darker region) and AA6061 in alternating layer configuration (Lee et al., 2003, Reprinted with permission from Springer).

Figure 4.3A illustrates the situation wherein A356 alloy was on the advancing side of the weld, whereas Figure 4.3B represents the case where AA6061 was present on the advancing side. Both the welds were made at the tool rotation rate of 1600 rpm and the tool traverse speed of 87 mm/min. The presence of onion ring in the nugget should be noted in both cases. Figure 4.3C and D shows optical micrographs taken from the nugget of the weld. An alternating layer of A356 (darker region) and AA6061 (brighter region) should be noted here in both cases.

The scanning electron microscopy images for as-cast A356 and wrought AA6061 alloys are shown in Figure 4.4A and B. A typical cast microstructure for A356 should be noted here in Figure 4.4A. The hardness profile for a weld corresponding to the case where AA6061 was on the advancing side is shown in Figure 4.4C. The inset of Figure 4.4C shows variation of the Si particles and Mg_2Si precipitates across different zones of the weld. The nugget of the weld shows the absence of Mg_2Si precipitates due to dissolution and coarsened Si particles. Note that the nugget hardness lies in between the base materials hardness. A356 alloy has a lower hardness than AA6061 alloy in base material.

In dissimilar aluminum alloys weld, the last two examples represented welds formed between cast–cast and cast–wrought combinations. Next

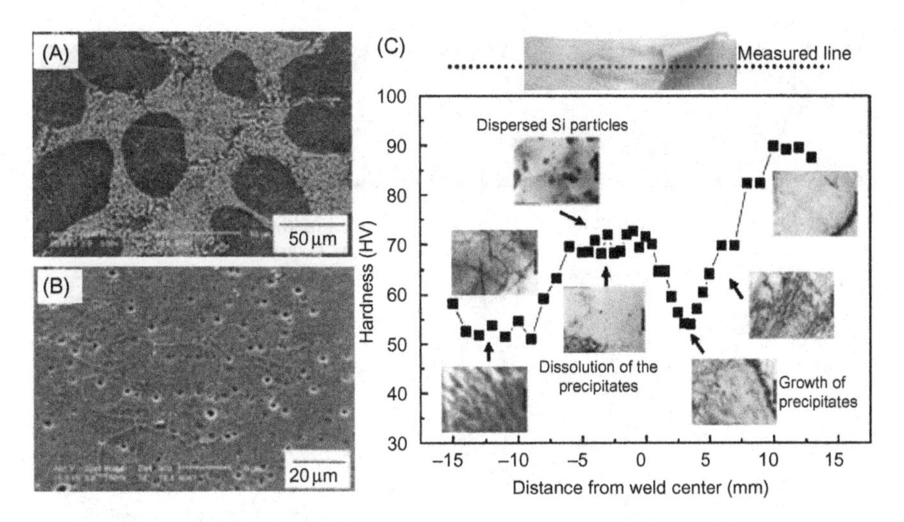

Figure 4.4 Scanning electron micrograph of (A) as-cast A356 and (B) wrought AA6061; (C) microhardness profile on the transverse cross-section of A356/AA6061 weld (AA6061 on advancing side). Insets in (C) show state of precipitates in different zones of the weld (Lee et al., 2003, Reprinted with permission from Springer).

we will discuss the weld formed between two wrought alloys. A transverse cross-section of the weld formed between AA6061-T6 (advancing side) and AA6082-T6 (retreating side) is shown in Figure 4.5. It shows different zones of the weld marked in approximate manner on the macrograph (Figure 4.5A). The nugget region shows that the materials from both alloys are present without much mixing. The lighter region represents AA6061 alloy. A micrograph which was taken from region 4 in Figure 4.5A is included in Figure 4.5B. Different grain sizes can be noted in the lighter (AA6061) and darker (AA6082) regions which are dispersed with very fine precipitates. Two-dimensional microhardness mapping of the nugget region revealed that there are two zones of maximum hardness along the depth of the nugget, and maximum hardness values were close to each other (Figure 4.5C). The microhardness measurement across different weld zones on the transverse cross-section at half the thickness of the plate revealed somewhat asymmetric variation of hardness across the weld centerline. The lowest hardness was found in the HAZ on retreating side (AA6082-T6 side). The hardness profile is shown in Figure 4.6. It also shows hardness profiles for similar welds AA6061 and AA6082 alloys.

So far the focus was to look at effect of starting processing conditions on material mixing, microstructural evolution, and mechanical properties of the weld. Now we discuss a few examples to show effects of other processing parameters such as tool rotation rates, tool traverse

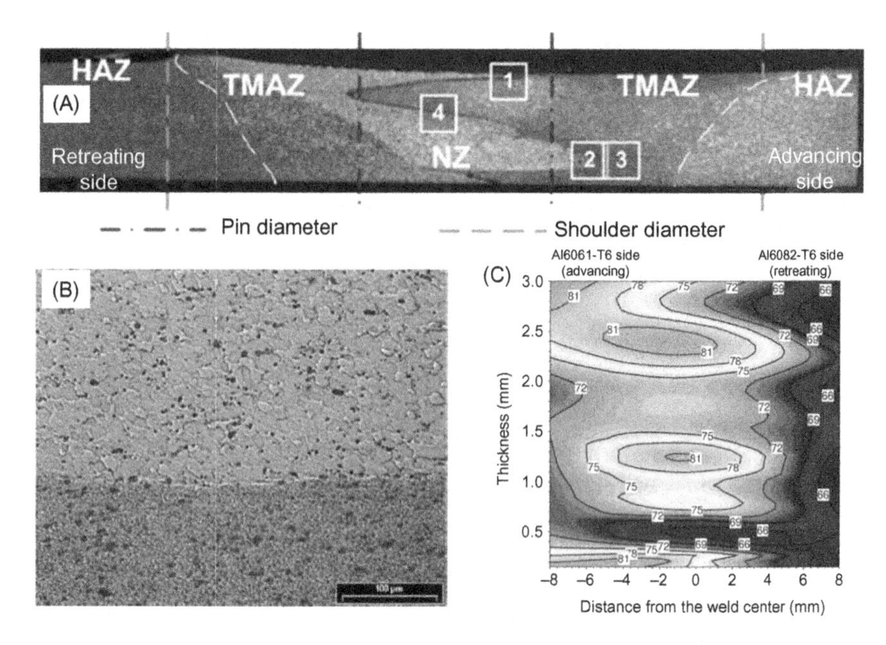

Figure 4.5 Dissimilar weld between AA6082-T6 (retreating side) and AA6061-T6 (advancing side). The weld was made at 1120 rpm tool rotation rate and 224 mm/min tool traverse speed. (A) Macrograph showing transverse cross-section of the weld, (B) micrograph corresponding to the region bounded by the rectangle labeled 4 in (A), and (C) 2D hardness contour map showing hardness of the material along the depth within the nugget zone of the welded structure (Moreira et al., 2009).

speeds, materials position with respect to the welding tool on parameters such as material mixing, microstructure, and mechanical properties.

Figure 4.7 shows transverse cross-section of the dissimilar friction stir welds made at different tool rotation rates and tool traverse speeds. The alloy 5J32 (Al–Mg–Cu)—an aluminum alloy designation of Kobe steel and is equivalent to AA5023—was placed on the retreating side of weld. AA5052 was placed on the advancing side of the weld. Both were welded in butt configuration. Tool rotation rates used during welding were 1000 and 1500 rpm. The tool traverse speed varied from 100 to 400 mm/min in the steps of 100 mm/min at each tool rotation rate. As can be noted from Figure 4.7, in the welded zone, there is a sharp interface between these two alloys indicating very little mixing between the alloys during material flow around the FSW tool. However, at 1500 rpm, one can notice vortex-like feature in the welded zone which is indicative of enhanced material mixing at this tool rotation rate. Hence, enhanced mixing is definitely a result of high heat input due to higher tool rotation rate.

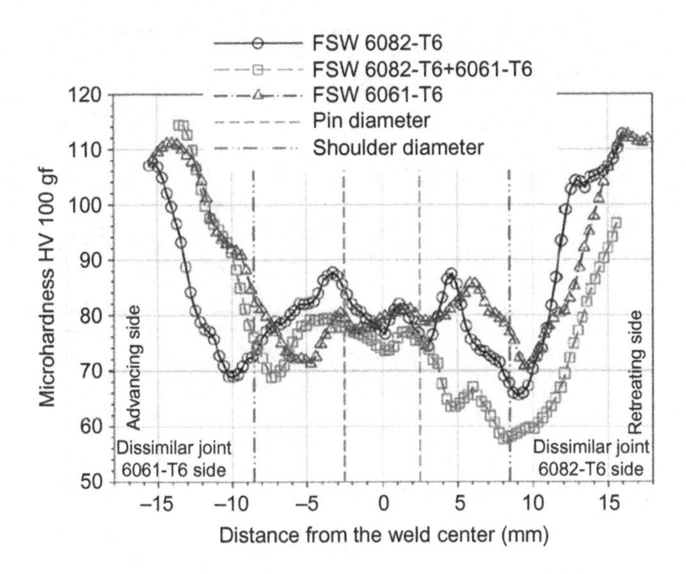

Figure 4.6 Microhardness profile across different zones on the transverse cross-section for the weld shown in Figure 4.5A along with two other similar welds (Moreira et al., 2009).

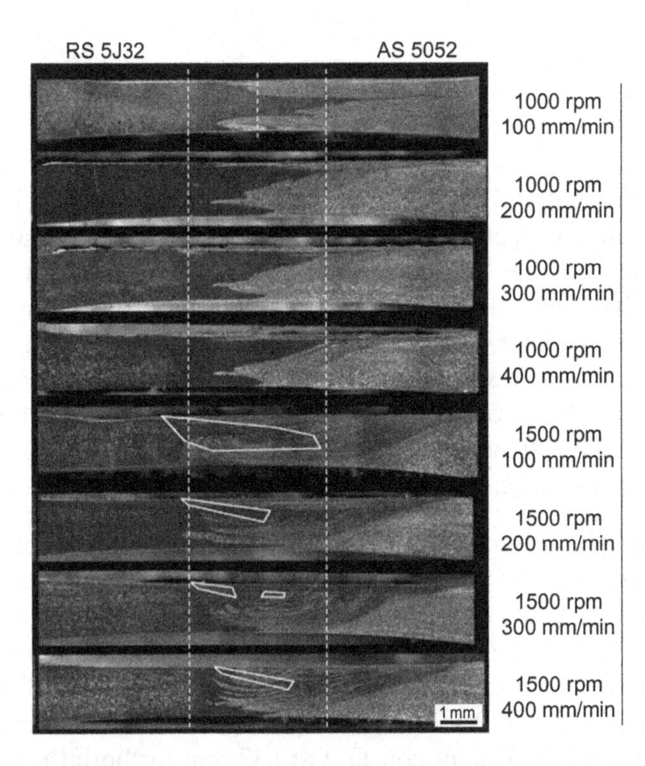

Figure 4.7 Effect of tool rotation rates and tool traverse speeds on the level of material mixing in 5J32 (retreating side) and AA5052 (advancing side) welds (Song et al., 2010, Reprinted with permission from The Japan Institute of Metals and Materials).

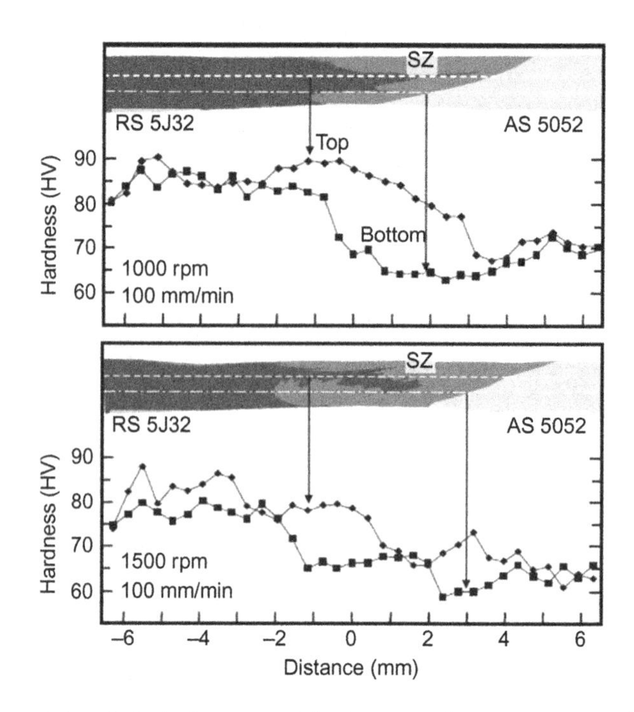

Figure 4.8 The variation of microhardness across different zones on the transverse cross-section of RS5J32-AS5052 dissimilar friction stir welds (Song et al., 2010, Reprinted with permission from The Japan Institute of Metals and Materials).

Enhanced material mixing is expected to have beneficial impact on the mechanical properties of the joints. Figure 4.8 shows variation of microhardness at the transverse cross-section of the weld for the welds made at 1000 rpm, 100 mm/min and 1500 rpm, 100 mm/min. Since there was inadequate material mixing between two alloys at 1000 rpm and 100 mm/min, it shows relatively sharp change in microhardness values across from 5J32 to AA5052. However, the weld made at 1500 rpm and 100 mm/min shows a gradual variation in hardness values from the retreating side to the advancing side.

Although Figure 4.7 shows welds made at different welding speeds, the effect of variation in tool traverse speed appears to be not very significant. To show the effect of tool traverse speed on material flow and mixing, another example is included in Figure 4.9. It shows the transverse cross-sectional images (macro- and microstructural) of the welds made at two different tool traverse speeds—57 and 229 mm/min. The tool rotation rate was kept constant at 637 rpm for both the welds and a tool with threaded pin was used for the welding. The dissimilar welds

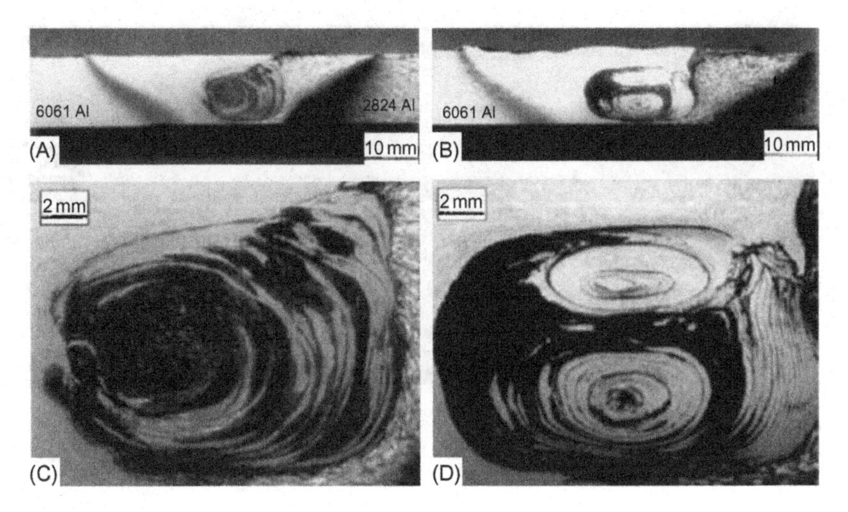

Figure 4.9 AA6061–AA2024 (advancing side) dissimilar welds made at 57 mm/min (A and C); 229 mm/min (B and D); tool rotation rate for both welds: 637 rpm (Ouyang and Kovacevic, 2002, Reprinted with permission from Springer).

were made between AA6061 and AA2024 alloys. Figure 4.9A and B shows the macroscopic images, and Figure 4.9C and D shows high-magnification images of weld nugget. For both welds, onion ring- like features should be noted in the weld nuggets. However, the weld made at 57 mm/min (Figure 4.9A and C) shows a better mixing and therefore less heterogeneity in the nugget compared to the weld made at 229 mm/min. For the weld made at 229 mm/min, in the nugget two onion rings- like features should be noted. Although the top onion ring- like feature consists of alternate layers of both alloys, based on the contrast (Figure 4.9D) it can be stated that it consisted of mostly AA6061 alloy. The onion ring toward the bottom of the plate also showed more of AA6061 alloy. In addition to this in the upper part of the weld, in the shoulder affected region, the amount of AA6061 alloy extruded toward advancing side of the weld appears to be more. It signifies less mixing at higher welding speeds.

To illustrate the role of workpiece position, whether advancing or retreating side, on material mixing and its effect on microstructural evolution and resulting mechanical properties, the transverse cross-sections of the welds made between AA5052-H32 and AA6061-T6 are shown in Figure 4.10. Both the welds were made at 2000 rpm tool rotation rate and 100 mm/min tool traverse speed. Figure 4.10A corresponds to the case where AA6061 alloy was placed on the

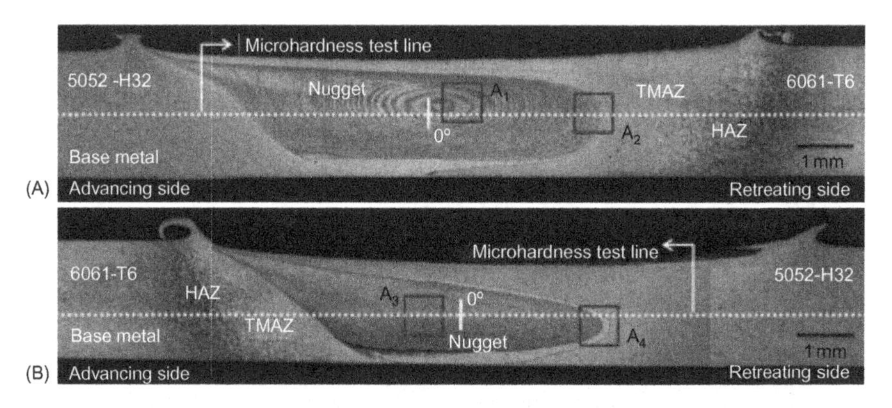

Figure 4.10 The effect of material positioning on the level of material mixing during friction stir dissimilar welding of 5052Al-H32 and 6061Al-T6 alloys (Park et al., 2010, Reprinted with permission from Maney Publishing).

retreating side of the weld, whereas Figure 4.10B is the case where AA6061 was present on the advancing side of the weld. The effect of the position of materials during welding is evident from the appearance of the welds in Figure 4.10. The onion ring in Figure 4.10A signifies proper mixing between AA5052 and AA6061 alloys during welding. However, no such feature is apparent in Figure 4.10B where AA6061 alloy was on the advancing side of the weld. The electron probe microanalysis study of the alternating bands in the onion ring of the weld in Figure 4.10A informed that Mg content in the dark-colored region was very close to that present in AA5052-H32, and in light-colored region, the Mg concentration was very close to that found in AA6062-T6 alloy. In the weld shown in Figure 4.10B Mg concentration did not vary much from point to point, and the average concentration was that of the AA6061-T6. Guo et al. (2014) also characterized the chemical compositions of various layers observed in the nugget in dissimilar weld made between AA6061 and AA7075 aluminum alloys. Figure 4.11 shows the energy dispersive spectroscopy (EDS) results from the work of Guo et al. (2014). Spectrum 1 shows that the material in the layer belongs to AA6061 alloy, spectrum 2 indicates the layer is made of AA7075 alloy, and spectrum 3 indicates that the layer is a mixture of AA6061 and AA7075. Hence, in the dissimilar welding, alternating layers represent different alloys used in the welding and their chemistry is dependent on level of intermixing. In this regard, the position of the alloy in dissimilar welding is very important which also affects the material mixing and overall composition of the welded region.

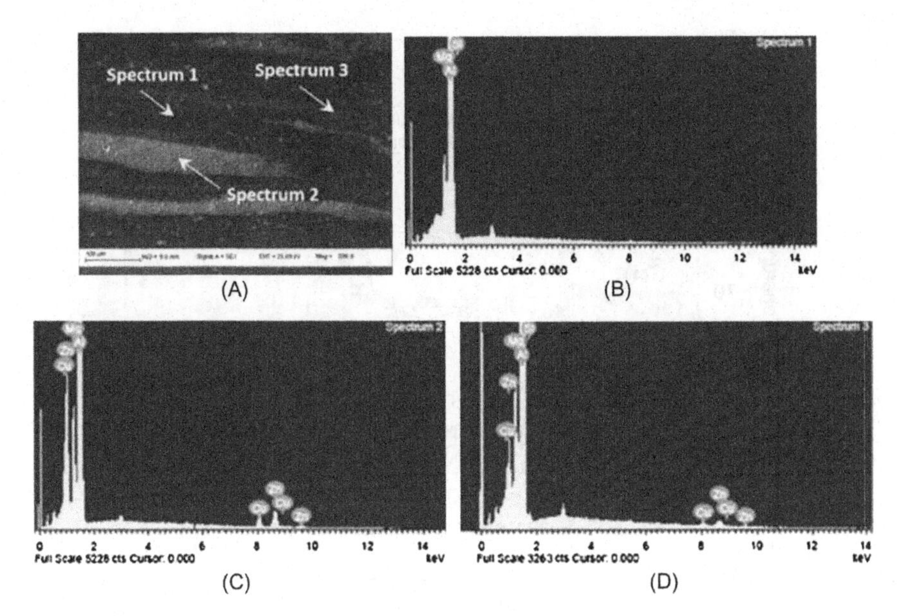

Figure 4.11 EDS analysis of the nugget region of the dissimilar weld between AA6061 and AA7075 alloys (Guo et al., 2014).

Differences in material flow are expected to manifest itself in different mechanical properties. The mechanical properties of the welds in Figure 4.10 were evaluated in terms of microhardness profiles for each weld on the transverse cross-section of the weld, and they are included in Figure 4.11. The microhardness profile in Figure 4.12A shows two local minima each corresponding to HAZ on each side. Between these two minima, the HAZ in AA5052 alloy on advancing side showed the minimum value. The minimum was still observed in the HAZ of AA5052 alloy when it was placed on the retreating side. However, for this configuration of the weld, the change in hardness values from nugget to HAZ was very sharp. Such a sudden change in microhardness values between different zones may have influence on deformation behavior of the welded structure.

The tool geometry also plays a very important role in the joint formation and quality of the joint formed. In a work along this line, Izadi et al. (2013) investigated effect of three different tool geometries on the level of material mixing. The schematic of the tools used (only two) are shown in Figure 4.13. The shoulder diameter and tool pin height were 10 and 4 mm, respectively, for each tool. They differed

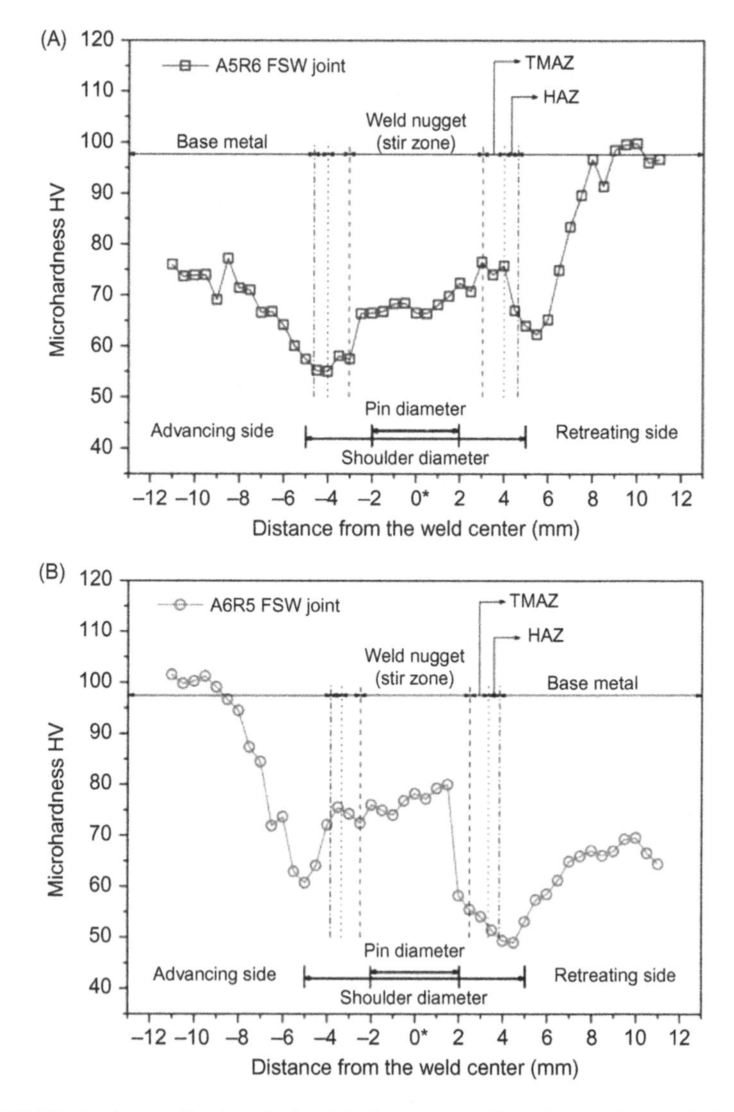

Figure 4.12 Microhardness profile along the dotted (yellow) horizontal lines shown in Figure 4.10 for the weld (A) A5R6 (5052Al-H32 on advancing side and 6061Al-T6 on the retreating side) and (B) A6R5 (6061Al-T6 on advancing side and 5052Al-H32 on the retreating side) (Park et al., 2010, Reprinted with permission from Maney Publishing).

from each other in terms of pin design. Two pin geometries were as follows—(i) pin having non-helical grooves with 0.7 mm spacing (Figure 4.13A) and (ii) pin with a geometry as in (i) with three flats on it, each apart by 120° (Figure 4.13B). All the welds were made at 894 rpm and 33 mm/min.

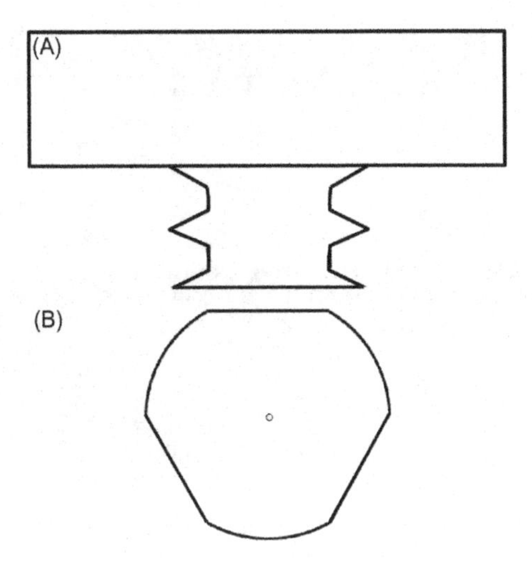

Figure 4.13 Schematic of the Tools with different pin geometries (Izadi et al., 2013).

Figure 4.14 shows the dissimilar welds made between AA2024 and AA6061 alloys using the tool shown in Figure 4.13A. Figure 4.14 also shows effect of workpiece position on the level of mixing between these two alloys. It is evident that in this case the level of intermixing remained the same despite the change in position of the alloys with respect to each other. The darker region of the weldments shown in Figure 4.14 represents AA2024 alloy. As shown in Figure 4.7 (for welds made at 1000 rpm and 100–400 mm/min) and Figure 4.10B, there is a sharp interface in the nugget between the materials belonging to these alloys. It shows very little intermixing between the alloys during FSW. The example included in Figure 4.14 indicates that by changing the position alone a better mixing between different materials cannot be guaranteed. Position of the material definitely has a key role to play in the level of mixing between different materials, but a suitable combination of other processing parameters such as tool rotation rates and tool traverse speeds are important in promoting the mixing of materials during dissimilar welding.

Figure 4.15 shows clearly the role tool geometry plays on materials mixing in the course of dissimilar material welding. When AA2024 was on the advancing side, the lower part of the nugget showed onion ring kind of structure in the weld made using tool pin having grooves and three flats on it. However, the top of the nugget shows a big chunk of AA6061 on retreating side extruded toward advancing side of the weld. The upper part is similar to what was observed in Figure 4.14A.

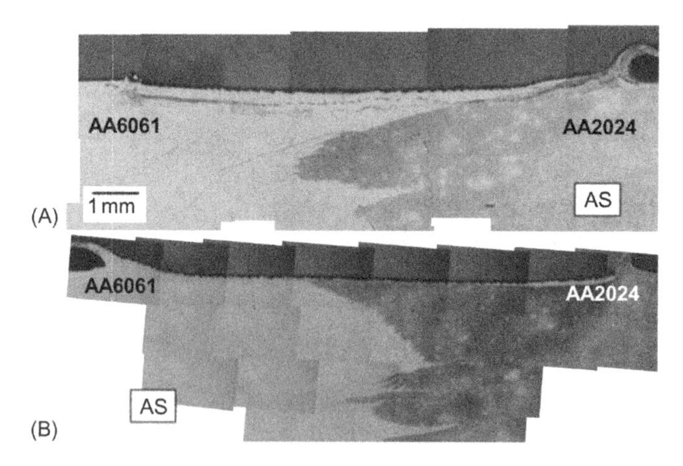

Figure 4.14 Friction stir dissimilar welding of AA6061 and AA2024 alloys using the tool pin with non-helical grove (Figure 4.13A) (Izadi et al., 2013, Reprinted with permission from Maney Publishing).

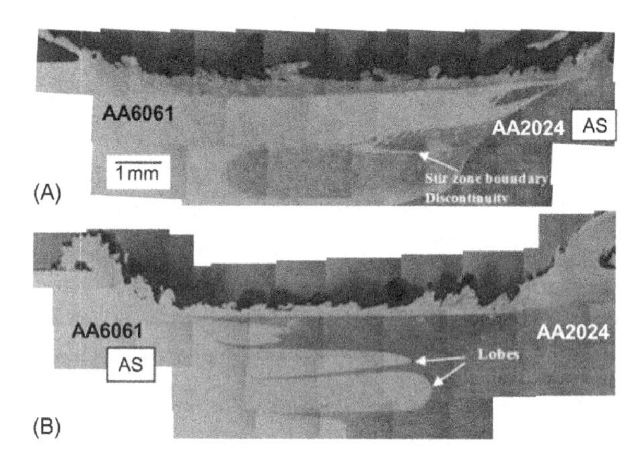

Figure 4.15 Friction stir dissimilar welding of AA6061 and AA2024 alloys using the tool pin with groove and three flats (Figure 4.13B) (Izadi et al., 2013, Reprinted with permission from Maney Publishing).

It appears that the flats on the pin promote vertical flow which results in a better mixing of different materials in dissimilar material welding. Due to the dominance of the shoulder in the upper part of the weld, the effect of flats on the pin is not evident. But, it is clearly evident from the onion ring in the lower part of the weld nugget in the pin-affected zone. When AA2024 was placed on the retreating side of the weld, it appears that the pin geometry does not enhance material mixing. Figure 4.15B shows that AA6061 (on advancing side) and AA2024 simply get extruded into the weld nugget without much intermixing (evident from the sharp interface). As shown in Figures 4.8 and 4.12, differences in material mixing will lead to different responses for a given loading condition in structural applications. Table 4.1 provides a summary of friction stir welding process parameter used in the welding of dissimilar aluminum alloys.

4.1.2 Steel to Steel

Although the technological importance of dissimilar steel welding cannot be overemphasized, the progress in this direction has been rather slow. The reasons for such a slow response have been mostly the same as for the FSW of similar welding of steels. Extensive tool wear and high cost of the tool materials have been a bottleneck for satisfactory growth for FSW of steels and other high-temperature materials.

Figure 4.16 shows dissimilar friction stir welds between low- and high-carbon steels. The low-carbon steel has been referred to as SPHC and high-carbon steels as SK85. The starting microstructure of SPHC mostly consisted of α-ferrite and that of SK85 was composed of ferritic matrix embedded with globular cementite. For making welds, a polycrystalline boron nitride (PCBN) tool having convex shoulder and tapered pin profile was used. The shoulder diameter, pin diameter at the shoulder–pin interface, and pin diameter at the tip of the pin were 25, 8, and 6 mm, respectively. The pin height was 3 mm. The dissimilar friction stir weld shown in Figure 4.16 was made at 800 rpm tool rotation rate and 200 mm/min in an argon atmosphere.

As shown before for dissimilar FSW of aluminum alloys, here also in Figure 4.16A, mixed and unmixed regions can be easily noted. As a result different regions in the nugget show different microstructures. For example, Figure 4.16C which corresponds to material present in shoulder-affected region; the microstructure consisted of fine pearlite and globular cementite particles. It indicates mixing between SPHC and SK85 to some extent. The micrograph in Figure 4.16D from region labeled "d" shows a microstructure similar to as-received SPHC. Figure 4.16E which belongs to region "e" shows 65% martensite. The microstructure of regions "f" and "g" consisted of pearlite + ferrite (Figure 4.16F) and ferrite + cementite + fine pearlite (Figure 4.16G), respectively. Different microstructures in various parts of the nugget indicate different level of mixing in different zones as a result of material flow and also different heating and cooling rates.

Table 4.1 Summary of Welding Parameters Used for Dissimilar FSW of Aluminum Alloys

Dissimilar Alloys	Tool Rotation Rate (rpm)	Tool Traverse Speed (mm/min)	Plate Dimension (mm)			Tool Material	Tool Dimensions			Tool Pin Profile	Reference
			W	*L*	*T*		*D*	*d*	*h*		
A356 and A7075	710, 1000, 1400	80, 112, 160	100	50	4	SKD11	20	4	3.8	Cylindrical, featureless	Boonchouytan et al. (2014)
AA6061–AA7075	1200	120, 180, 300	300	50			15	5 at root		Conical, threaded	Guo et al. (2014)
AA2024–AA7075	1500	50–300	200	100	5		15		6	Concave shoulder, conical, threaded	Song et al. (2010) (lap joint)
AA6061–AA2014	500, 1500	90	550	70	4.7		15	5	4.4	Scrolled shoulder, threaded cylindrical pin with three flats	Jonckheere et al. (2013)
AA2024–AA6061	894	33–88 (88 for lap welding)			6.35 (AA6061 1 mm thick for lap welding)	H13	10	4		Cylindrical, threads with no helix, helical threads, threads with three flats 120° apart	Izadi et al. (2013)
AA1100-B_4C–AA6063	2000	100, 200	150	50	4.4	WC-Co				Conical featureless	Guo et al. (2012)
AA5083–AA6351	600, 950, 1300	60	100	50	6	High-carbon, high-chromium steel	18	6	5.7	Cylindrical with 4, 6, and 8 flats, conical with 4 and 8 flats	Palanivel et al. (2012)
AA5086–AA6061	600–1000	30–150	150	50	5	H13	20	6 at the root; 3 at the tip	4.8	Conical with a conical slot (2° angle) on the surface of the pin	Jamshidi Aval et al. (2012)

AA2024–AA7075	400, 1000, 2000	254			3		12	4	2.85	Cylindrical threaded	da Silva et al. (2011)
A356–AA6061	1000, 1400	80–240	100	30	3	High-speed steel	15	5	2.6	Cylindrical	Ghosh et al. (2010)
AA2024–AA7075	400, 1200, 1500	100, 150, 400			1.2, 2.0, 2.5		12	3			Zadpoor et al. (2010)
A356–A319	1120, 1400, 1800	80, 112	250	50	10	H13	35	5	8	Cylindrical	Hassan et al. (2010)
AA5052–AA5J32 (Al–Mg–Cu)	1000, 1500	100–400					12	3.8	1.45	Cylindrical threaded	Song et al. (2010)
AA5052–AA6061	2000	100			2		10	4	1.7	Concave shoulder, cylindrical pin	Park et al. (2010)
AA2017–AA6005	1000	200	600	70	6		18	8	5.7	Cylindrical threaded with three flutes	Simar et al. (2010)
AA6061–AA6082	1120	224			3		17			Cylindrical threaded pin with concave shoulder	Moreira et al. (2009)
AA2024–AA6056	500–1200	150–400			4		15	5		Cylindrical threaded pin with concave shoulder	Amancio-Filho et al. (2008)
AA2024–AA7075	1200	42–200			3	SKD61	12	4		Cylindrical threaded	Khodir and Shibayanagi (2008)
AA2024–AA7075	1600	120	200	80	4						Cavaliere and Panella (2008)
A356–AA6061	1600	87–267	140	70	4						Lee et al. (2003)

(*Continued*)

Table 4.1 (Continued)

Dissimilar Alloys	Tool Rotation Rate (rpm)	Tool Traverse Speed (mm/min)	Plate Dimension (mm)			Tool Material	Tool Dimensions			Tool Pin Profile	Reference
			W	*L*	*T*		*D*	*d*	*h*		
AA5083–AA6061	890, 1540	118, 155			3		10	3	2.8	Cylindrical	Shigematsu et al. (2003)
AA2024–AA6061	151–914	57–330			12.7	Tool steel					Ouyang and Kovacevic (2002)
AA2024–AA6061	400–1200	60			6.5	Carbon steel	19	6.5			Li et al. (1999)

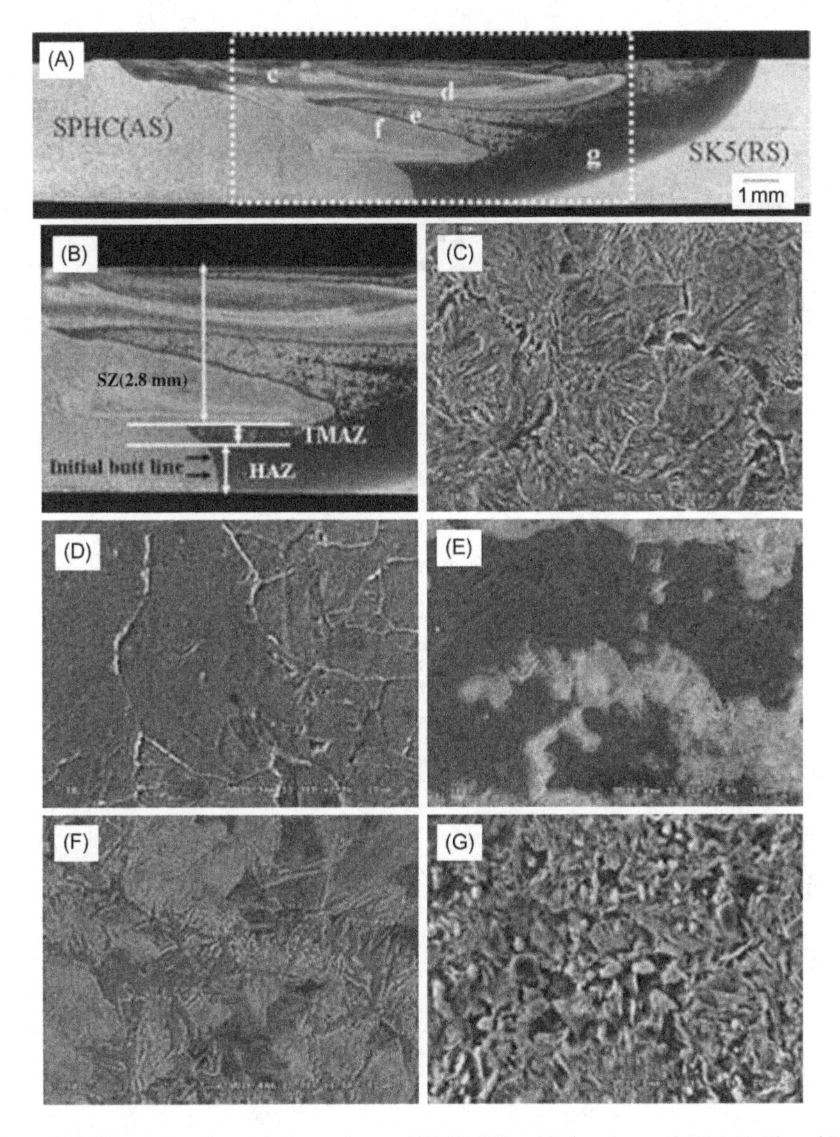

Figure 4.16 FSW of dissimilar steels: low-carbon steel (0.04 wt%) and high-carbon steel (0.84%). The welding was carried out at 800 rpm and 200 mm/min in argon atmosphere using a PCBN tool (Choi et al., 2011, Reprinted with permission from The Japan Institute of Metals and Materials).

Figure 4.17 shows mechanical properties for the dissimilar weld for which microstructural results were presented in Figure 4.16 along with another weld made at 400 rpm tool rotation rate. Microhardness profiles are shown in Figure 4.17A, and tensile test results are included in Figure 4.17B. As can be noted at both tool rotation rates, the welded

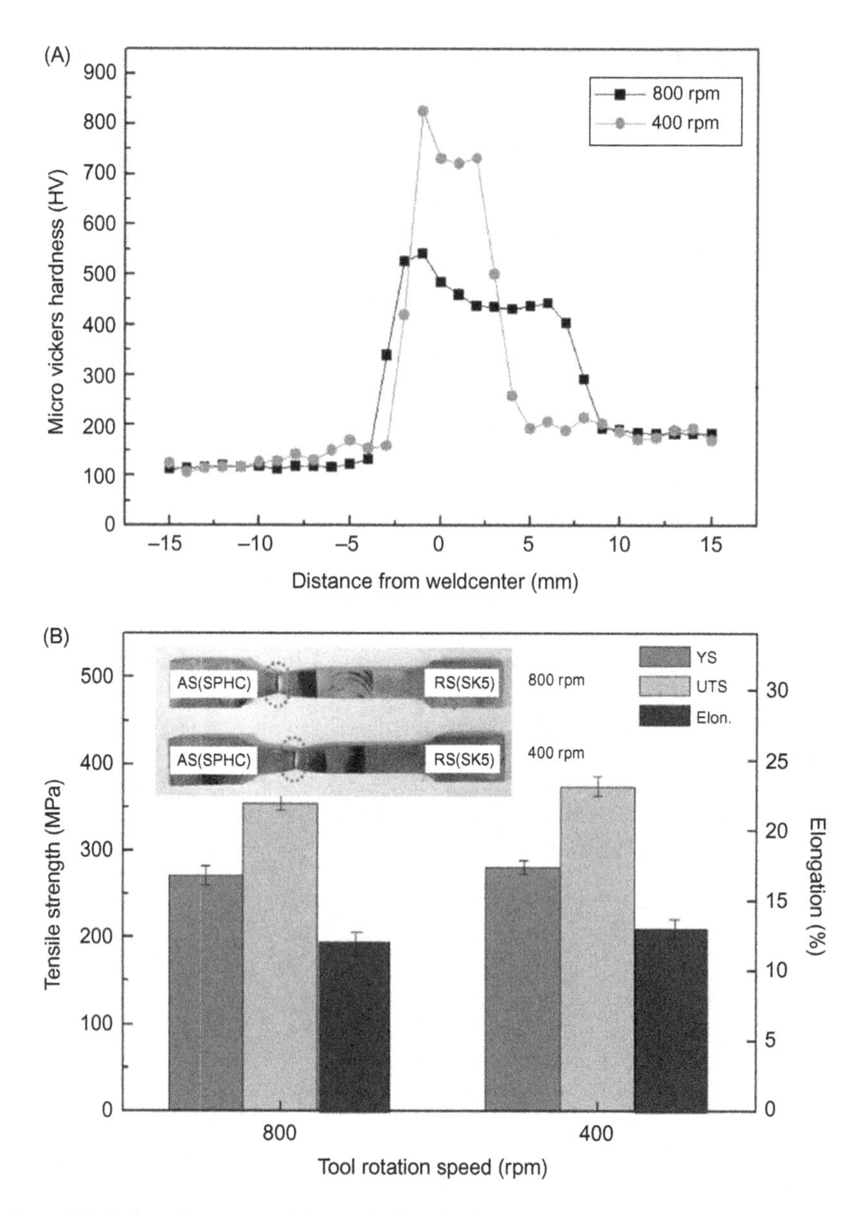

Figure 4.17 Mechanical properties. (A) Microhardness results; measurement was carried out on the transverse cross-section and (B) tensile test results and inset showing fractured tensile samples. The microstructural results were presented in Figure 4.15 for the weld made at 800 rpm. The mechanical properties result also include results for the weld made at 400 rpm and welding parameters same as for 800 rpm weld (Choi et al., 2011, Reprinted with permission from The Japan Institute of Metals and Materials).

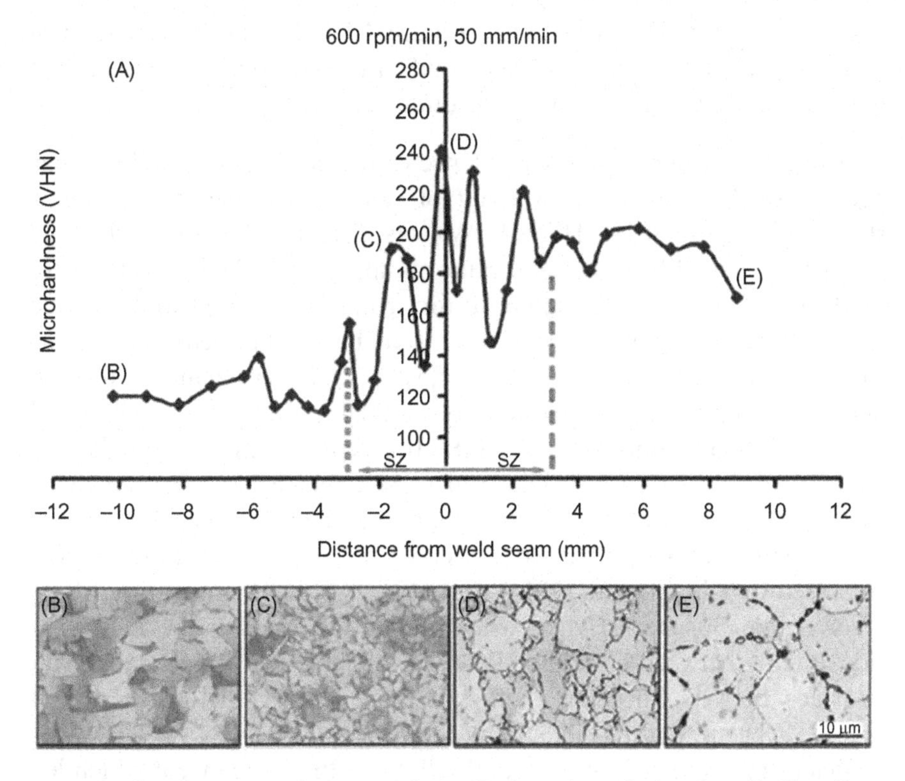

Figure 4.18 Effect of microstructural variation on the microhardness values in the dissimilar welds. The weld was made between low-carbon steel (0.11 wt%C) and stainless steel 304. The tool was made of WC-Co and the tool dimensions were as follows: tool shoulder diameter—16 mm, pin diameter—5.5 mm, and pin height—2.6 mm (Jafarzadegan et al., 2013).

zones (nugget, TMAZ, and HAZ) show hardness values higher than the base materials. From Figure 4.17B, it should be noted that tensile tests results show almost the same level of YS, UTS, and %El for both tool rotation rates. The inset in Figure 4.17B shows the location of fracture of tensile samples. Clearly, the fracture took place outside the welded zone.

Although the microhardness profile show very little variation in the values in the nugget in this case, the variation can be quite significant due to highly heterogeneous microstructure as a result of differences in material flow characteristics and different heating and cooling rates in the different parts of the nugget. Figure 4.18 shows microhardness profiles for dissimilar weld between stainless steel and low-carbon steel. Overall in the stir zone the hardness increases from low-carbon steel side to stainless steel side. But within the stir zone a significant variation in microhardness values should be noted in Figure 4.18. The variation is as high as 100 VHN. Figure 4.18 also shows micrographs from the different regions of the weld, and positional dependence of microstructure within the nugget is quite evident here.

The effect of positional dependence of materials during welding has been studied in dissimilar welding of steels also. One such result is shown in Figure 4.19. Here FSW was performed between ferritic/martensitic steel F82H and austenitic stainless steel 304. All welds were made at 100 rpm and 100 mm/min tool rotation rate and tool traverse speeds, respectively, using WC-based tool. The tool dimensions were as follows: shoulder diameter—15 mm, pin diameter—6 mm, and pin length—1.3 mm. In this welding, the vertical surface of the cylindrical pin was shifted 0.1 mm toward F82H. It means that the tool pin did not penetrate in the 304 steel. The transverse cross-sections for the welds show the differences in materials mixing in each case in the weld nugget. It was explained on the basis of softening behavior of the alloys. It was discussed by Chung et al. (2011) that at the welding temperature F82H is softer than 304 steel. Hence, when F82H was placed on the advancing side, mixing between F82H and 304 was not sufficient which leads to an interface between these two alloys (Figure 4.19(A)). However, when placed on the retreating side, temperature toward the 304 steel side was sufficient to cause softening in both alloys to the same extent which led to better mixing. As explained in Chapter 2, it may also have to do with interface position with respect to the tool rotation axis. From material mixing point of view, it is advisable to position the interface biased toward advancing side (refer to Chapter 2 for the details).

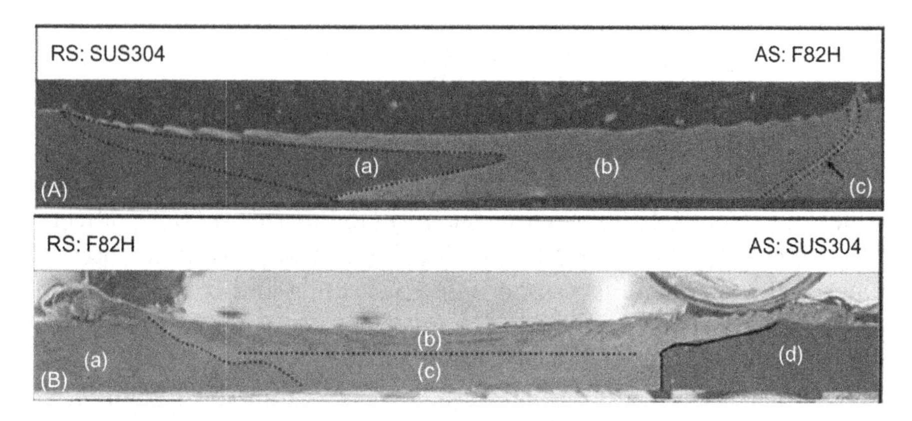

Figure 4.19 Effect of alloy position on materials mixing in dissimilar FSW of ferritic/martensitic F82H and austenitic 304 steel (Chung et al., 2011).

4.2 FRICTION STIR LAP WELDING OF DISSIMILAR ALLOYS

This section describes a few examples based on dissimilar friction stir lap welding. In the lap welding, two or more workpieces are stacked on top of each other. Like the butt welding, there are a number of parameters which affect joint integrity in the lap welding. Figure 4.20 shows stacking order of AA5052 and AA6061 alloys sheets used in the dissimilar friction stir lap welding by Lee et al. (2008). The thickness of AA6061-T6 and AA5052-H112 alloys were 2 and 1 mm, respectively. In addition to the effect of workpiece staking order, Lee et al. (2008) also studied the effect of tool rotation rate and tool traverse speed on material mixing and resulting mechanical properties.

Figure 4.21 shows the transverse cross-section of welds made at different tool rotation rates. The configuration of workpieces corresponds to that shown in Figure 4.20A. In Figure 4.21A–C the welds were made by varying the tool rotation rates and in Figure 4.21D–F by varying the tool traverse speeds. Irrespective of the processing parameters, the top layer of the welded zone consisted of only AA5052. Onion ring patterns should also be noted which are developed to different degrees depending on the processing parameters. As discussed for the butt welding earlier, here also the alternating layers in the onion ring pattern correspond to the alloys used in the welding. The presence of AA5052 alloy in the lower part of the nugget suggests vertical flow of the material during the welding, and the extent of this flow is definitely dependent on the processing parameters. For example, Figure 4.21C shows an almost complete onion ring compared to the one shown in Figure 4.21A and B. The tool rotation rate was highest for Figure 4.21C and lowest for Figure 4.21A. Friction stir welds shown in Figure 4.21D–F were

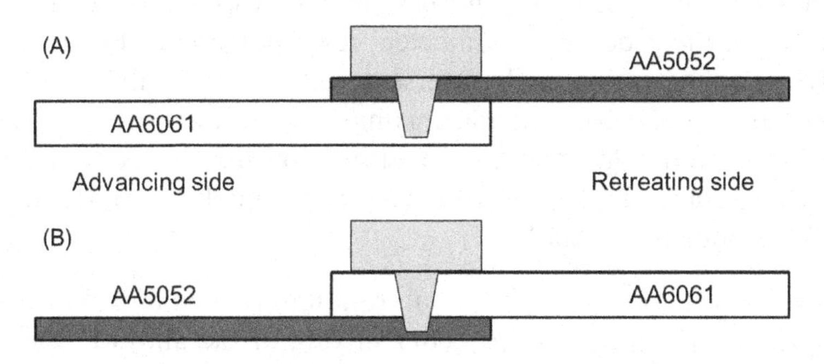

Figure 4.20 Schematic showing different lap weld configuration (A) top sheet AA5052 and (B) AA6061 alloys.

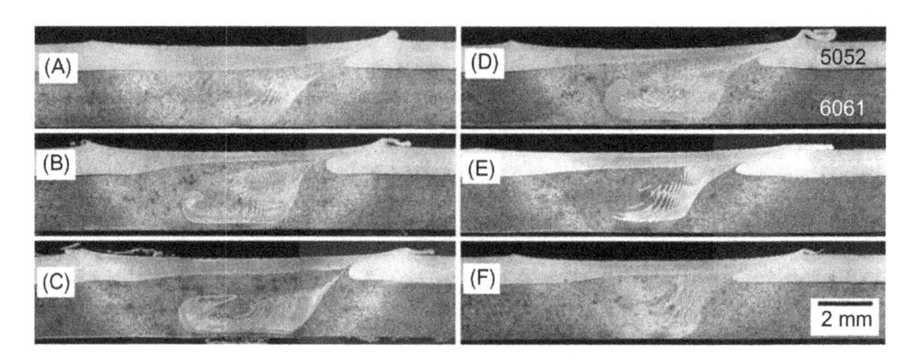

Figure 4.21 Friction stir dissimilar lap welds made between AA5052 and AA6061 by (A–C) increasing tool rotation rate and (D–F) increasing tool traverse speed. In both cases, the lap weld configuration corresponded to Figure 4.21A (Lee et al., 2008, Reprinted with permission from Springer).

made at different tool traverse speeds while keeping other processing parameters constant. Note that the contrast in the lower part of the weld in Figure 4.21D signifies the presence of both alloys due to better mixing, which disappears gradually on increasing the tool traverse speed. The weld shown in Figure 4.21F corresponds to the highest tool traverse speed, and it shows that the bottom part of the nugget predominantly consists of one material (AA6061). It can be rationalized based on less time for material to flow vertically and mix at higher tool traverse speeds.

Figure 4.22 has been included to illustrate the effect of the alloy position in the lap welding. All the welds shown in Figure 4.22 were made by placing AA5052 sheet below AA6061. Overall, a relatively low level of mixing (compared to the results shown in Figure 4.21) between AA6061 and AA5052 alloys is evident from the transverse cross-sectional images presented in Figure 4.22. Among the welds shown in Figure 4.22A–C, the best in terms of material mixing is observed for the processing parameters corresponding to Figure 4.22C. Similarly, at increasing tool traverse speed material mixing becomes less effective. The poor material mixing is definitely a result of lower temperature and relatively smaller strain toward the bottom of the weld sheet stacking. It causes AA5052 to simply extrude and get embedded into the AA6061.

Note that although higher tool rotation rate and/or lower tool traverse speed is necessary for better mixing of the alloys in the weld nugget, a better mixing does not necessarily translate into better load

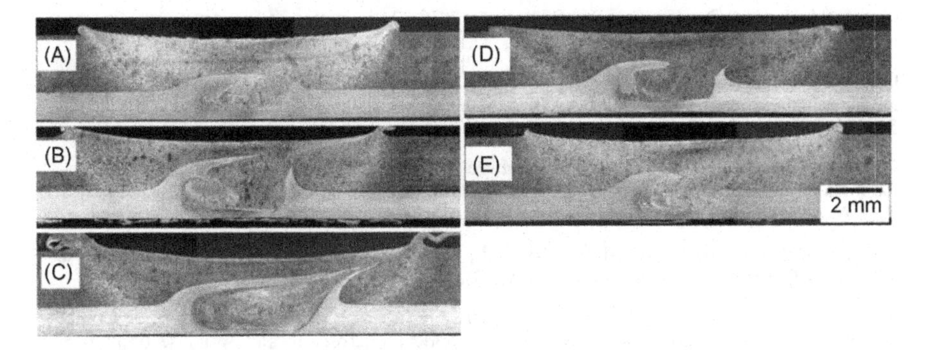

Figure 4.22 Friction stir dissimilar lap welds made between AA5052 and AA6061 by (A–C) increasing tool rotation rate and (D and E) increasing tool traverse speed. In both cases, the lap weld configuration corresponded to Figure 4.20B (Lee et al., 2008, Reprinted with permission from Springer).

carrying capacity of the joints. Note that higher tool rotation rate and/or lower tool traverse speed is also related to higher heat input during the welding. It might result in inferior mechanical properties in welded zones. For example, the lap shear test carried out by Lee et al. (2008) for the weld corresponding to 1250 rpm and 267 mm/min (Figure 4.22A) fractured at 5.6 kN, whereas the weld made at 2500 rpm and 267 mm/min (Figure 4.22B) showed a load carrying capacity of 2.2 kN. Hence, it suggests that process parameter optimization to obtain best combinations of material mixing and mechanical properties.

REFERENCES

Amancio-Filho, S.T., Sheikhi, S., Dos Santos, J.F., Bolfarini, C., 2008. Preliminary study on the microstructure and mechanical properties of dissimilar friction stir welds in aircraft aluminium alloys 2024-T351 and 6056-T4. J. Mater. Process. Technol. 206 (1), 132–142.

Boonchouytan, W., Chatthong, J., Rawangwong, S., Romadorn, R., Muangjunburee, P., 2014. Investigation of dissimilar joint between 356 and 7075 of semi-solid (SSM) aluminum alloy by friction stir welding. Adv. Mater. Res. 931–932, 344–348.

Cavaliere, P., Panella, F., 2008. Effect of tool position on the fatigue properties of dissimilar 2024-7075 sheets joined by friction stir welding. J. Mater. Process. Technol. 206, 249–255.

Choi, D.H., Ahn, B.W., Yeon, Y.M., Park, S.H.C., Sato, Y.S., Kokawa, H., et al., 2011. Microstructural characterizations following friction stir welding of dissimilar alloys of low- and high-carbon steels. Mater. Trans. 52, 1500–1505.

Chung, Y.D., Fujii, H., Sun, Y., Tanigawa, H., 2011. Interface microstructure evolution of dissimilar friction stir butt welded F82H steel and SUS304. Mater. Sci. Eng. A 528, 5812–5821.

da Silva, A.A.M., Arruti, E., Janeiro, G., Aldanondo, E., Alvarez, P., Echeverria, A., 2011. Material flow and mechanical behaviour of dissimilar AA2024-T3 and AA7075-T6 aluminium alloys friction stir welds. Mater. Des. 32, 2021–2027.

Ghosh, M., Kumar, K., Kailas, S.V., Ray, A.K., 2010. Optimization of friction stir welding parameters for dissimilar aluminum alloys. Mater. Des. 31, 3033–3037.

Guo, J.F., Chen, H.C., Sun, C.N., Bi, G., Sun, Z., Wei, J., 2014. Friction stir welding of dissimilar materials between AA6061 and AA7075 Al alloys effects of process parameters. Mater. Des. 56, 185–192.

Guo, J., Gougeon, P., Chen, X.-G., 2012. Microstructure evolution and mechanical properties of dissimilar friction stir welded joints between AA1100-B4C MMC and AA6063 alloy. Mater. Sci. Eng. A 553, 149–156.

Hassan, A.S., Mahmoud, T.S., Mahmoud, F.H., Khalifa, T.A., 2010. Friction stir welding of dissimilar A319 and A356 aluminium cast alloys. Sci. Technol. Weld. Join. 15, 414–422.

Izadi, H., Fallu, J., Abdel-Gwad, A., Liyanage, T., Gerlich, A.P., 2013. Analysis of tool geometry in dissimilar Al alloy friction stir welds using optical microscopy and serial sectioning. Sci. Technol. Weld. Join. 18, 307–313.

Jafarzadegan, M., Abdollah-zadeh, A., Feng, A.H., Saeid, T., Shen, J., Assadi, H., 2013. Microstructure and mechanical properties of a dissimilar friction stir weld between austenitic stainless steel and low carbon steel. J. Mater. Sci. Technol. 29, 367–372.

Jamshidi Aval, H., Serajzadeh, S., Kokabi, A.H., 2012. Experimental and theoretical evaluations of thermal histories and residual stresses in dissimilar friction stir welding of AA5086-AA6061. Int. J. Adv. Manuf. Technol. 61, 149–160.

Jonckheere, C., de Meester, B., Denquin, A., Simar, A., 2013. Torque, temperature and hardening precipitation evolution in dissimilar friction stir welds between 6061-T6 and 2014-T6 aluminum alloys. J. Mater. Process. Technol. 213, 826–837.

Khodir, S.A., Shibayanagi, T., 2008. Friction stir welding of dissimilar AA2024 and AA7075 aluminum alloys. Mater. Sci. Eng. B 148, 82–87.

Lee, W.B., Yeon, Y.M., Jung, S.B., 2003. The mechanical properties related to the dominant microstructure in the weld zone of dissimilar formed Al alloy joints by friction stir welding. J. Mater. Sci. 38, 4183–4191.

Lee, C.-Y., Lee, W.-B., Kim, J.-W., Choi, D.-H., Yeon, Y.-M., Jung, S.-B., 2008. Lap joint properties of FSWed dissimilar formed 5052 Al and 6061 Al alloys with different thickness. J. Mater. Sci. 43, 3296–3304.

Li, Y., Murr, L.E., McClure, J.C., 1999. Solid-state flow visualization in the friction-stir welding of 2024 Al to 6061 Al. Scr. Mater. 40, 1041–1046.

Moreira, P.M.G.P., Santos, T., Tavares, S.M.O., Richter-Trummer, V., Vilaça, P., de Castro, P.M.S.T., 2009. Mechanical and metallurgical characterization of friction stir welding joints of AA6061-T6 with AA6082-T6. Mater. Des. 30, 180–187.

Ouyang, J.H., Kovacevic, R., 2002. Material flow and microstructure in the friction stir butt welds of the same and dissimilar aluminum alloys. J. Mater. Eng. Perform. 11, 51–63.

Palanivel, R., Mathews, P.K., Murugan, N., Dinaharan, I., 2012. Effect of tool rotational speed and pin profile on microstructure and tensile strength of dissimilar friction stir welded AA5083-H111 and AA6351-T6 aluminum alloys. Mater. Des. 40, 7–16.

Park, S.K., Hong, S.T., Park, J.H., Park, K.Y., Kwon, Y.J., Son, H.J., 2010. Effect of material locations on properties of friction stir welding joints of dissimilar aluminium alloys. Sci. Technol. Weld. Join. 15, 331–336.

Shigematsu, I., Kwon, Y.-J., Suzuki, K., Imai, T., Saito, N., 2003. Joining of 5083 and 6061 aluminum alloys by friction stir welding. J. Mater. Sci. Lett. 22, 353–356.

Simar, A., Jonckheere, C., Deplus, K., Pardoen, T., de Meester, B., 2010. Comparing similar and dissimilar friction stir welds of 2017–6005A aluminium alloys. Sci. Technol. Weld. Join. 15, 254–259.

Song, S.-W., Kim, B.-C., Yoon, T.-J., Kim, N.-K., Kim, I.-B., Kang, C.-Y., 2010. Effect of welding parameters on weld formation and mechanical properties in dissimilar Al alloy joints by FSW. Mater. Trans. 51, 1319–1325.

Zadpoor, A.A., Sinke, J., Benedictus, R., 2010. Global and local mechanical properties and microstructure of friction stir welds with dissimilar materials and/or thicknesses. Metall. Mater. Trans. A 41, 3365–3378.

CHAPTER 5

Friction Stir Welding of Dissimilar Materials

5.1 Al TO Mg ALLOYS

Welding of dissimilar Al and Mg alloys has been paid increasing attention recently due to their lightweight characters and high specific strengths for structural applications in transportation industries. Al and Mg alloys have very similar melting points, not quite different coefficient of thermal expansion and thermal conductivity, but different formability due to different crystal structures. Welding of dissimilar Al and Mg alloys is challenging due to the formation of liquation and brittle intermetallic compounds (IMCs), $Mg_{17}Al_{12}$ and Al_3Mg_2 through diffusion process. As shown in the Al–Mg phase diagram (Figure 5.1), two eutectic reactions can occur at much lower temperatures compared to the melting temperatures. Even with friction stir welding (FSW), the peak temperature in the stir zone is generally above these eutectic points. Welding between Al and Mg using FSW typically leads to the formation and growth of IMC layer at the interface. $Mg_{17}Al_{12}$ and Al_3Mg_2 are typically present in the IMC layers which can be in various morphologies depending on the alloy systems and welding conditions adopted. Due to the brittle characters of these IMCs and easy formation of cracks in the IMC layers, strength of the welds is always degraded when a thicker IMC layer is formed accompanied by the solidification cracking. In order to increase the weld strength, reducing the Al–Mg IMC reaction layer thickness as well as the solidification cracking is necessary.

Among available literature on FSW of Al to Mg alloys, regardless the use of various alloy systems, welding parameters, or configurations, one common feature is the formation of IMC at the interface between Al and Mg alloys. Figure 5.2 is just one example showing high-magnification view at interface IMC layer between Mg alloy AZ31 and Al alloy AA5083. The welds were produced in a butt configuration with Al alloy on the advancing side. A tool rotation rate of 500 rpm was employed in Figure 5.2A and 450 rpm for Figure 5.2B

Friction Stir Welding of Dissimilar Alloys and Materials. DOI: http://dx.doi.org/10.1016/B978-0-12-802418-8.00005-9

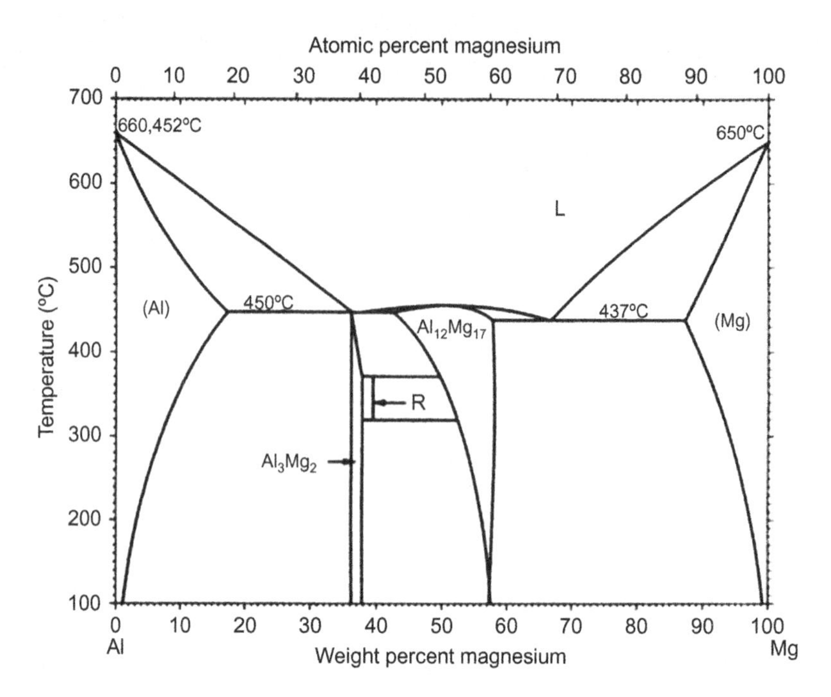

Figure 5.1 The binary Al–Mg phase diagram (Murray, 1992).

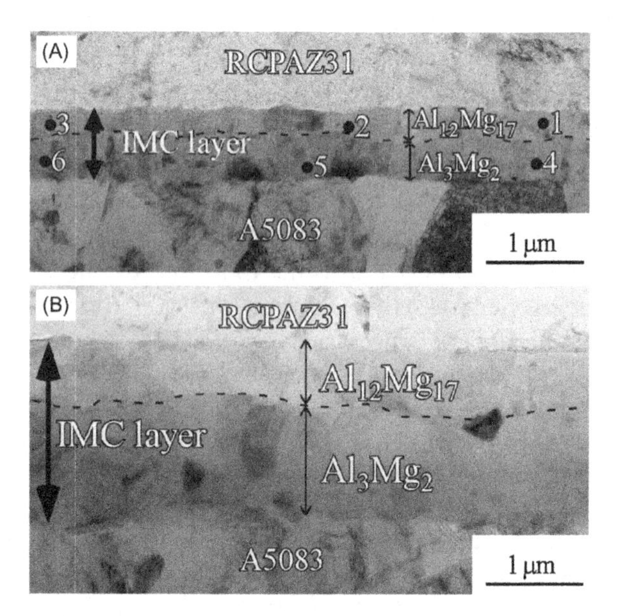

Figure 5.2 Transmission electron microscope (TEM) images showing formation of IMCs at the interface between Mg alloy AZ31 and Al alloy AA5083 in a butt welding configuration (A) without tool offset and (B) with tool offset of 1.5 mm to Al alloy (Yamamoto et al., 2009, Reprinted with permission from The Japan Institute of Metals and Materials).

with same tool traverse speed of 100 mm/min. In both cases, IMCs including $Al_{12}Mg_{17}$ and Al_3Mg_2 are distributed regularly within the IMC layer which is about 0.8 μm thick for the former and 1.8 μm thick for the latter. The difference in thickness of IMC layer is related to the different frictional heat inputs due to tool offset, which has been discussed later in this section. Comparing these two figures, one can also realize that the thickness of $Al_{12}Mg_{17}$ is less than that of Al_3Mg_2 in both welding conditions, which indicates different growth kinetics for these two IMCs.

The formation and growth of Al–Mg IMCs can be induced by diffusion. Figure 5.3 presents the computed thickness of IMC layer as a function of diffusion time for the case shown in Figure 5.2A based on an assumption of a diffusion control for IMC formation and growth, which was later verified by the experimental observation. The activation energy for Al_3Mg_2 is less than that of $Al_{12}Mg_{17}$, as a result, the growth rate for Al_3Mg_2 is much faster. It is noted that the IMC layer grows quickly and the thickness of the layer is over 1 μm within 4 s at 573 K. The TEM observation indicates fine grain structure for both IMCs without any indication of solidification features.

As specified in the Al–Mg phase diagram, there are two low-melting-temperature eutectic reactions, which can take place at significantly lower temperatures as compared to the melting temperature of either metals. However, these reaction temperatures are within the

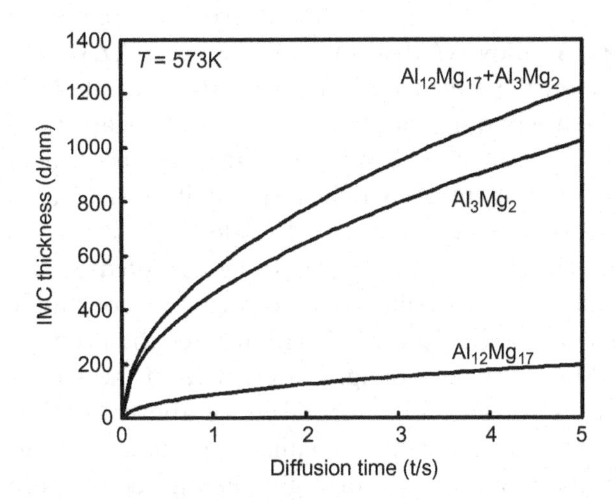

Figure 5.3 Thickness of IMC as a function of discussion time (Yamamoto et al., 2009, Reprinted with permission from The Japan Institute of Metals and Materials).

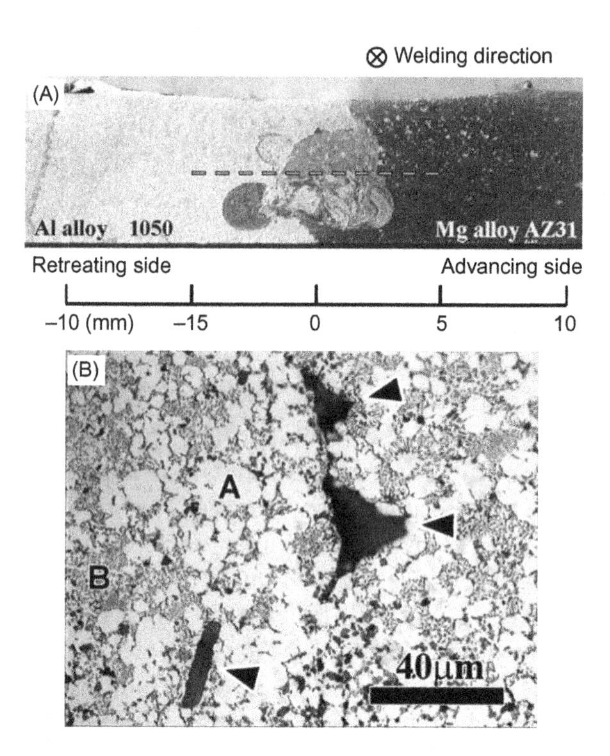

Figure 5.4 Friction stir-welded AA1050 and AZ31 in a butt configuration: (A) cross-sectional view of the welded interface and (B) microstructure of the interface showing solidified structure (Sato et al., 2004).

range of peak temperature during FSW, which indicates the occurrence of liquation and solidification structure in Al–Mg welding. Figure 5.4 presents a transverse cross section and microstructure of dissimilar butt welded 6 mm Al alloy AA1050 to same thickness AZ31 Mg alloy with the Mg alloy on the advancing side. The weld was made at 2450 rpm and 90 mm/min without tool offset. An irregular shaped region is visible at the weld center. The microstructure of the irregular shaped region shown in Figure 5.4B indicates solidified structure. Porosities and eutectic structure with black and white contrast are noticed. This eutectic structure was further verified to be consisted of $Al_{12}Mg_{17}$ and Mg solid solution. The result was interpreted as constitutional liquation during welding and eutectic reaction after the primary solidification of $Al_{12}Mg_{17}$ during cooling. Different from the results shown in Figure 5.2, a higher welding temperature of above 733 K is suggested in the weld region in this case. An enhanced mutual diffusion between Al and Mg atoms as a result of high welding temperature and high strain rate plastic deformation result in the constitutional liquation.

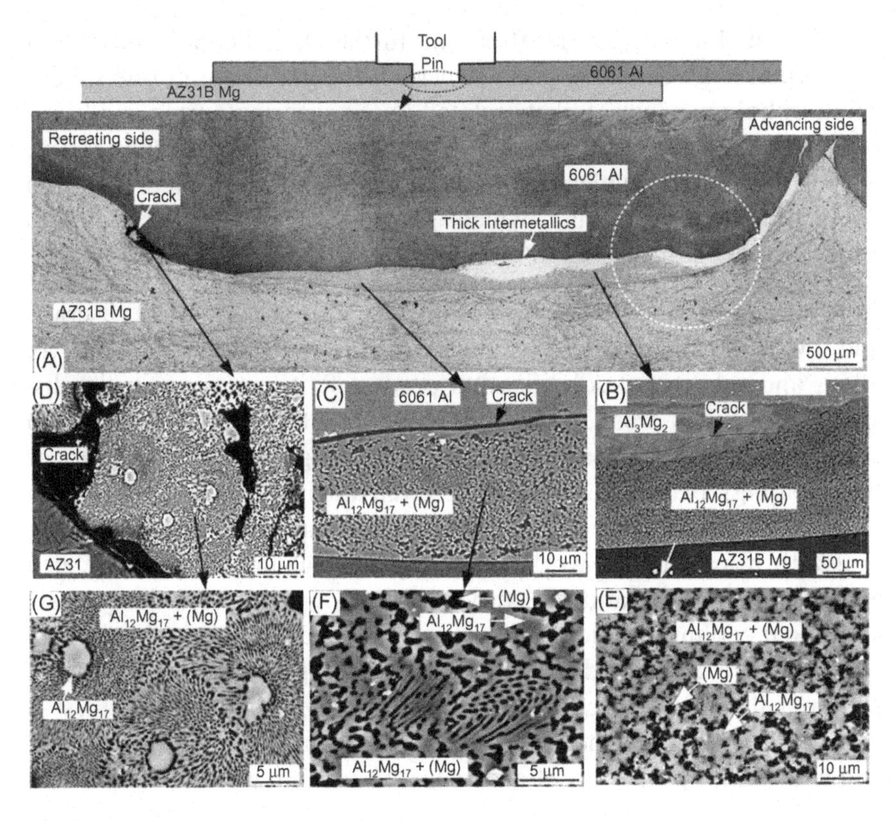

Figure 5.5 Friction stir lap welding of Al 6061 to AZ31: (A) macrostructure of transverse cross section and (B–G) Scanning electron microscope (SEM) microstructures at various locations (Firouzdor and Kou, 2010a), Reprinted with permission from Springer.

The formation of $Al_{12}Mg_{17}$ and eutectic structure is linked to the local chemical inhomogeneity which has 30–40 at.% of Al.

The interdiffusion can be more prominent when Al and Mg alloys are welded in a lap configuration due to larger contact area and the thermal history profile the weld experiences. This is generally represented by large area of IMCs. Figure 5.5 shows one example in this class where IMC formation through eutectic reaction takes place. In this particular case, Al alloy AA6061 with a thickness of 1.6 mm was friction stir welded to a 1.6 mm Mg alloy AZ31 in a lap configuration. The weld was made at a tool rotation rate of 1400 rpm and a tool traverse speed of 38 mm/min with a tool having 1.5 mm pin length. IMC layers formed at the bottom of the nugget between Al and Mg. In addition to the large Al_3Mg_2 layer adjacent to the upper Al alloy, a thick layer of laminar eutectic structure is shown as a result of constitutional liquation formation and further solidification. Instead of the normal large dendritic grain structure, the grains are in fine spherical morphology which is likely related to the shearing during solidification. Cracking was also noticed in the weld on the retreating side, serving as another evidence of liquation. Liquid is sheared into thin liquid films during welding, which in turn can cause cracking due to strain mismatch during solidification.

As discussed above, the formation and morphology of both crack and IMCs are related to the mutual diffusion between Al and Mg alloys and subsequent solidification process. Their presence in the dissimilar welds can weaken the welds significantly. Both diffusion and solidification processes are affected by welding configuration (lap or butt welding), placement of workpiece (advancing vs. retreating side in butt welding and upper vs. lower position in lap welding) and welding tool with respect to the workpiece interface, welding parameters, such as tool rotation rate and tool traverse speed, welding tool design, and additional cooling solution. Since the majority of the dissimilar Al and Mg FSW are conducted in butt welding configuration, the following section will mainly discuss some of the above mentioned factors in butt welding to show how they affect material flow, microstructure, and mechanical properties of dissimilar Al and Mg welds.

The placement of workpiece has significant influence on possibility to achieve sound dissimilar welds due to the differences in flow behavior of Al and Mg alloys and asymmetric temperature distribution on advancing and retreating sides. Table 5.1 summarizes dissimilar welding of Al alloys to AZ31, which is the most frequently used Mg alloy in dissimilar welding with Al alloys. It can be noted that the placement of workpiece for achieving desired sound welds varies among different Al alloy series. For welding 5XXX Al alloys to AZ31, Al alloy is most often placed on the advancing side; however, in terms of 6XXX Al alloys, Mg alloy on the advancing side is preferred. In some cases, investigation indicates improved weldability and performance with proper placement of workpieces.

The placement of workpiece influences the frictional heat input, and in turn, the material flow and intermixing. Figure 5.6 shows weld cross sections of butt-welded AA6061 to AZ31 with either Al or Mg on the advancing side. The welds were made without tool offset and using the same welding parameters at 1400 rpm tool rotation rate and

Table 5.1 Summary of Dissimilar Welding of Mg Alloy AZ31 to Various Al Alloys in Butt Welding

Al Alloy	Advancing Side	Tool Offset	Welding Parameters	Reference
Al 1050	Mg better	Unspecified	2450 rpm, 90 mm/min	Sato et al. (2004)
Al 1060	Al	Yes	200–1000 rpm, 19–75 mm/min	Yan et al. (2005)
Al 2024	Al	No	2500 rpm, 200–550 mm/min	Khodir and Shibayanagi (2007)
Al 2024	Al	Unspecified	Unspecified	Cao and Jahazi (2010)
Al 5083	Al	No	300–400 rpm, 60–100 mm/min	McLean et al. (2003)
Al 5052	Al	No	800–1600 rpm, 300 mm/min	Kwon et al. (2008)
Al 5052	Al	No	800–1600 rpm, 100–400 mm/min	Morishige et al. (2008)
Al 5052	Al	No	1000–1400 rpm, 100–500 mm/min	Shigematsu et al. (2009)
Al 5083	Al	Yes	300–600 rpm, 40–120 mm/min	Yamamoto et al. (2009)
Al 5052	Al	Unspecified	600 rpm, 40 mm/min	Yan et al. (2010)
Al 5754	Al better	Unspecified	600–1400 rpm, 40–300 mm/min	Simoncini and Forcellese (2012)
Al 5083	Al	Unspecified	400 rpm, 50 mm/min	Mofid et al. (2014)
Al 6040	Mg better	No	1400 rpm, 200–300 mm/min	Zettler et al. (2006)
Al 6061	Mg better	Yes	800, 1400 rpm, 38 mm/min	Firouzdor and Kou (2009)
Al 6040	Mg	No	1400 rpm, 225 mm/min	Kostka et al. (2009)
Al 6061	Mg	Unspecified	850 rpm, 48 mm/min	Chang et al. (2010)
Al 6061	Mg	Unspecified	400 rpm, 20 mm/min	Malarvizhi and Balasubramanian (2012)
Al 6061	Mg better	Unspecified	1500–2000 rpm, 80–200 mm/min	Bergmann et al., 2013
Al 6061	Mg	Yes	600–1000 rpm, 100 mm/min	Liang et al. (2013)
Al 6061	Mg	Yes	800 rpm, 5 mm/min	Regev et al. (2014)
Al 6061	Al	No	600–1400 rpm, 20–60 mm/min	Masoudian et al. (2014)
AC4C	Al	Unspecified	1500 rpm, 20–80 mm/min	Chen and Nakata (2008)

38 mm/min tool traverse speed. Figure 5.6A shows the weld having Al alloy on the advancing side with layers of Mg alloy carried into the stir zone and formation of a thin layer of IMC at the interface between AZ31 and AA6061. The stir zone was shifted toward the Al side. Figure 5.6B presents the cross section for the weld having Mg on the advancing side. A relatively even distributed stir zone with respect to the weld interface is observed in this case. Particularly, the Mg alloy penetrates deeply into the stir zone and interlocks with the Al alloy. Distinct material flow behaviors and microstructural evolution are argued to be related to differences in heat input for both

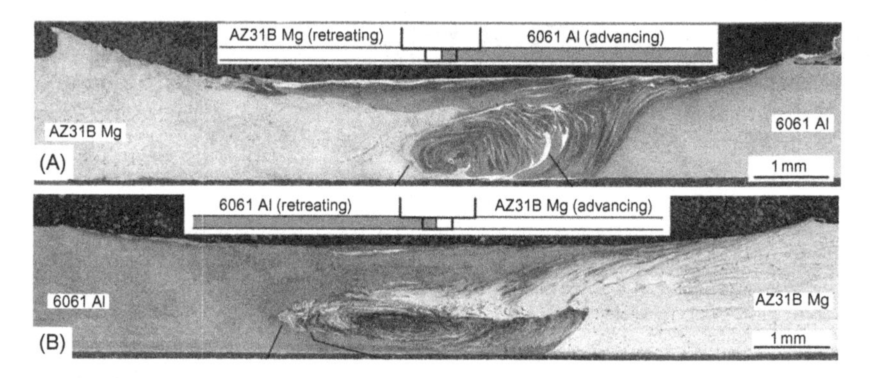

Figure 5.6 Cross sections of butt-welded AA6061 to AZ31 with (A) Al on the advancing side and (B) Mg on the advancing side (Firouzdor and Kou, 2010b, Reprinted with permission from Springer).

configurations. The heat input was claimed to be higher with Al on the advancing side than with Mg on the advancing side. Having Mg on the advancing side can reduce the heat input and the formation of IMC, thus improved weld integrity.

Placement of the tool pin plays an important role in material flow and intermixing. Figure 5.7 shows the transverse cross sections of butt-welded 3 mm AA6061 and an Mg alloy with 1.02 wt% Al and 1.05 wt.% Zn under a tool rotation rate of 900 rpm and a tool traverse speed of 100 mm/min. The Mg alloy was placed on the advancing side and tool was plunged either at the faying line without pin offset (Figure 5.7C) or with 1.5 mm offset to Al (Figure 5.7B) or Mg (Figure 5.7A) side. The welded regions show alternative gray and dark banded structures which were identified as containing intermetallic particles due to material mixing. A good amount of material mixing can be noticed without tool offset. With tool offset to the Mg side, less material mixing with reduced banded structure is observed. The welded zone is mainly filled with Mg alloy. With tool offset to the Al side, least material mixing is observed with no obvious banded structure formation. The differences in material mixing and band structure with respect to the tool offset are likely related to the frictional heat generation and local composition variation.

Different material mixing, formation, and distribution of IMCs in the weld zone lead to the difference in mechanical behavior of welds. Figure 5.8 shows engineering stress–strain curves of welds shown in Figure 5.7 under uniaxial tensile loading. The welds made without tool

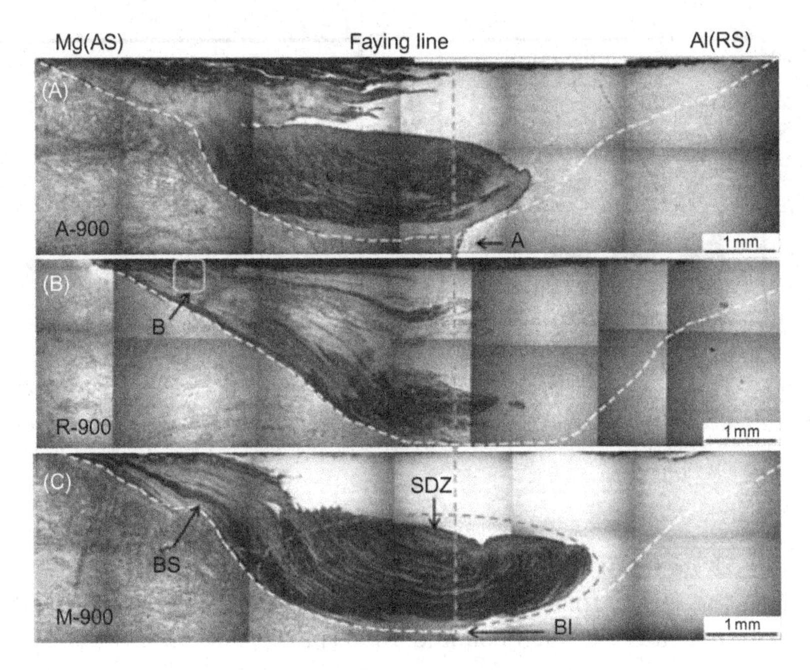

Figure 5.7 Transverse cross sections of butt-welded AA6061 and Mg–1.02 wt% Al–1.05 wt% Zn under various tool pin plunge positions, (A) 1 mm offset into the Mg side, (B) 1 mm offset into the Al side, and (C) without offset (Liang et al., 2013, Reprinted with permission from Springer).

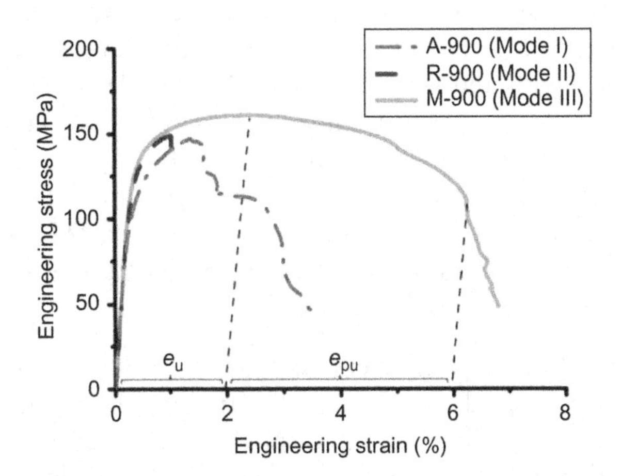

Figure 5.8 Effect of tool offset on the tensile behavior of joints shown in Figure 5.7 (Liang et al., 2013, Reprinted with permission from Springer).

offset (M-900) exhibits slightly higher tensile strength, but dramatically larger elongation to failure. It is noteworthy that welds without tool offset and with 1.5 mm tool offset to the Mg side failed from the intermixing region, and the weld with tool offset to the Al side failed within

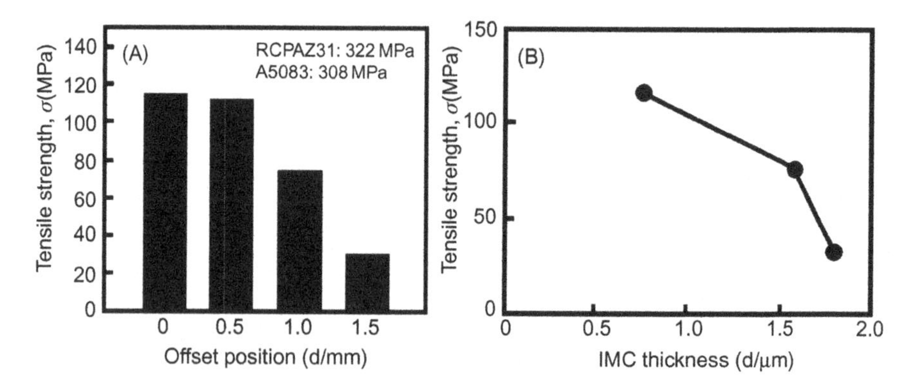

Figure 5.9 Tensile strength of dissimilar butt-welded AA5083 and AZ31 as a function of (A) tool offset and (B) the thickness of IMC layer (Yamamoto et al., 2009, Reprinted with permission from The Japan Institute of Metals and Materials).

the mixed material region in the nugget. The differences in tensile strength for this particular alloy system have been observed for other tool rotation rates between 600 and 1000 rpm.

Similar behavior on tensile strength versus tool offset has been observed for dissimilar butt welding of Al alloy AA5083 and AZ31 with Al on the advancing side and tool offset to Al up to 1.5 mm as shown in Figure 5.2. Figure 5.9A presents the tensile strength summary for welds with and without offsets. The tensile strength decreases significantly when an offset of 1 mm or above is used. The difference was interpreted as increased thickness of IMC layer and weakened mechanical interlocking effect. Figure 5.9B shows the relation between the tensile strength of welds and the thickness of IMC layer. It is noticed that the tensile strength decreases considerably when the thickness of the IMC layer is above 1.8 μm.

Generally, welding parameters such as tool rotation rate and traverse speed are coupled to influence the frictional heat generation and volume of material movement and disposition for a unit distance (ratio of tool rotation rate to tool traverse speed). Discussing one or the other separately may not show direct impact on process design when dissimilar Al and Mg welding is required, but can provide a general view on how each of them affect heat generation, material intermixing, microstructure, and mechanical properties. Figure 5.10 shows the effect of tool rotation rate on material mixing between 2 mm AA5052 and AZ31 in a butt configuration with Al on the advancing side. No tool

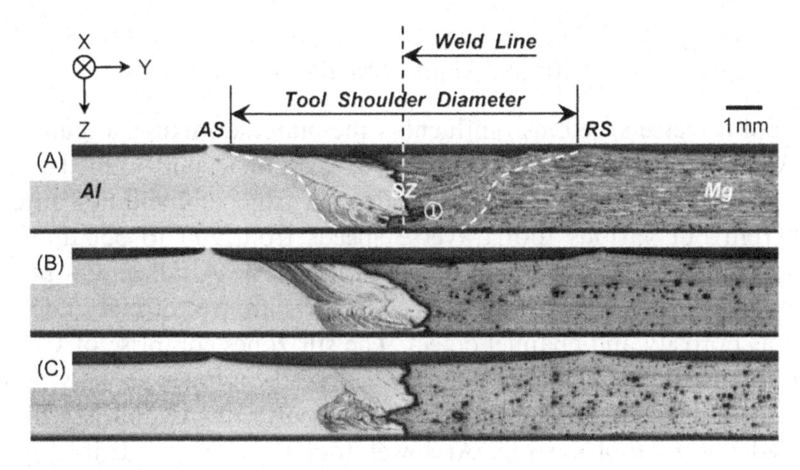

Figure 5.10 Transverse cross sections for dissimilar butt welds between Al 5052 and AZ31 under various tool rotation rates: (A) 1000 rpm, (B) 1200 rpm, and (C) 1400 rpm (Kwon et al., 2008).

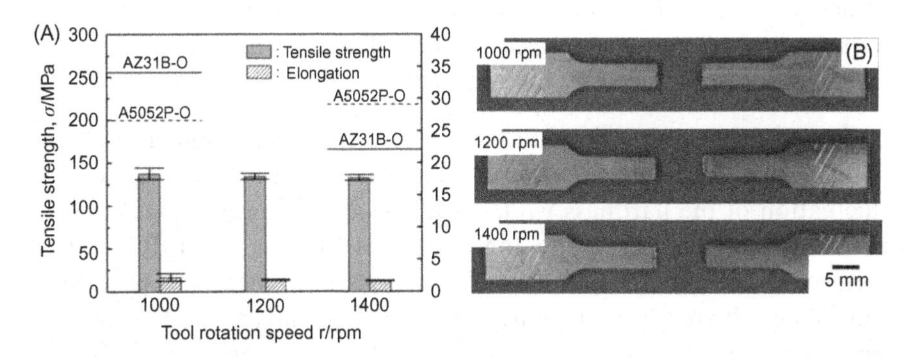

Figure 5.11 Tensile behavior of butt-welded Al 5052 and AZ31: (A) tensile strength and elongation as a function of tool rotation rate and (B) fracture behavior of welds shown in Figure 5.10 (Shigematsu et al., 2009, Reprinted with permission from The Japan Institute of Metals and Materials).

offset was used in this study. The welds were conducted under the same tool traverse speed of 300 mm/min and various tool rotation rates from 800 to 1600 rpm, and only those made between 1000 and 1400 rpm are defect-free. For welds made at 1000–1400 rpm, all bonded interface show similar and limited material mixing as evident as zigzag morphology in each case.

Figure 5.11 presents the effect of tool rotation rate on tensile strength for welds shown in Figure 5.10. There is no obvious variation in tensile strength for those three welds. This is also the case for the elongation. As shown in Figure 5.11B by the same authors, all welds fractured near the center region without much reduction of area near the fracture region. The fracture behavior is believed to be related to the abrupt interface between Al and Mg alloys.

Tool traverse speed also influences the material mixing. Figure 5.12 shows the transverse cross sections of dissimilar butt-welded 3 mm AA2024 and AZ31. Welds were made at a fixed tool rotation rate of 2500 rpm but various tool traverse speeds from 200 to 550 mm/min using a threaded pin tool without pin offset. The Al alloy was placed on the advancing side. All weld cross sections are free of macro defects such as porosity and channel defect. The stir zones comprise of various regions based on the different contrasts. These contrasts actually represent the lamellar band structure of mixing of both Al and Mg alloys formed due to tool stirring. At lower tool traverse speed, it appears that more material mixing occurs than that at higher tool traverse speed, likely due to the long period of stirring action of material in same volume, that is, longer residence time. Higher frictional heat induced by lower traverse speed is also believed to influence the material mixing as evidenced by the reduced stir zone size.

Figure 5.13 shows microhardness profiles of welds made at 200 and 550 mm/min shown in Figure 5.12. The stir zone exhibits a significant fluctuation of the hardness values, with relatively lower values close to the Mg side. The relatively higher values are due to IMCs which formed at the interface as well as the location where severe mixing took place. Increase in the traverse speed does affect the base metal, especially the Al alloy side; however, it does not appear to influence the stir zone much due to the fact that IMC formation happens in both cases.

As presented above, the formation of Al–Mg IMCs is mainly driven by interdiffusion of the species and is highly dependent on the specific thermal history (time and temperature) of the welding process. Placement of workpiece and the tool with respect to the weld interface can affect the IMC significantly, and in turn, the mechanical properties. Even though the tool geometry effect is not discussed specially, it can be noted from the examples mentioned before, the tool geometry can influence the material flow significantly. Featured pin such as threads and flats are believed to enhance material mixing. In addition, efforts have been made to reduce welding peak temperatures by controlling the frictional heat input as well as introducing media such as water or liquid nitrogen to quickly take away the generated heat, or

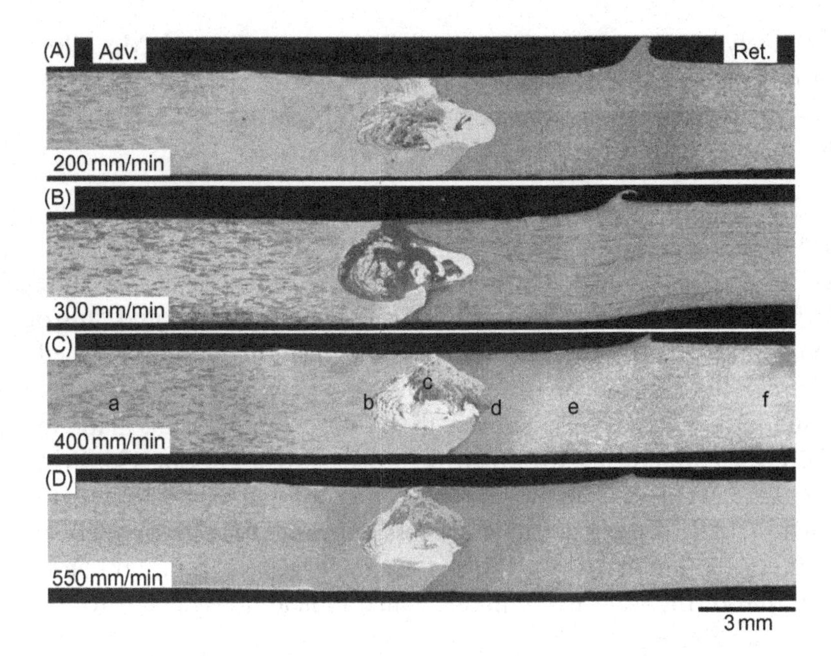

Figure 5.12 Transverse cross sections for butt-welded Al 2024 to AZ31 under 2500 rpm and various tool traverse speeds: (A) 200 mm/min, (B) 300 mm/min, (C) 400 mm/min, and (D) 5500 mm/min (Khodir and Shibayanagi, 2007, Reprinted with permission from The Japan Institute of Metals and Materials).

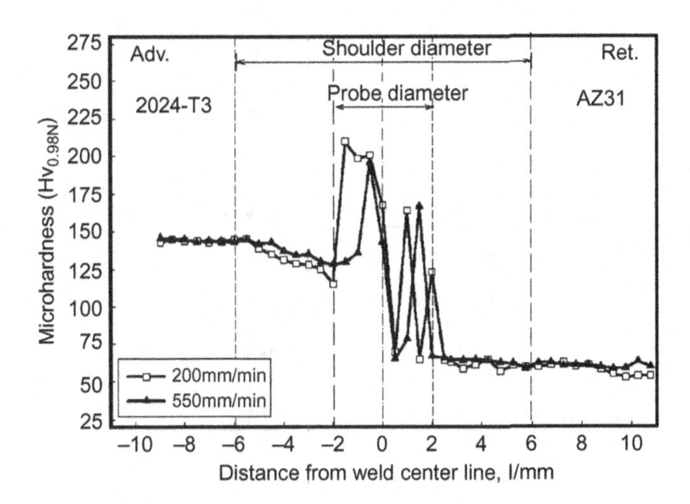

Figure 5.13 Microhardness profile for dissimilar butt welds between AA2024 and AZ31 under 200 and 550 mm/min (Khodir and Shibayanagi, 2007, Reprinted with permission from The Japan Institute of Metals and Materials).

selectively placing a third material in between Al and Mg alloys to form IMC which is stronger but less brittle as compared to the Al–Mg IMCs. The later endeavors have been discussed at the end of this chapter with other dissimilar alloy systems.

One example of how cooling can affect the IMC formation and growth is shown in Figure 5.14. In this particular case, AA5083 and AZ31 were butt welded using the same parameters but in different environments; in air, submerged in water and liquid nitrogen. Temperature profiles for particular locations are specified in Figure 5.14A. It can be observed that the peak temperature obtained in submerged butt welds is lower than that of the welding made in air. The peak temperature dropped by 46–53°C from about 435°C in air, when the weld was made in water or liquid nitrogen, respectively. Figure 5.15 presents the hardness profile for welds shown in Figure 5.14. The hardness values for submerged welds are much lower than that of the weld made in air, which indicates less IMCs formation in the stir zone.

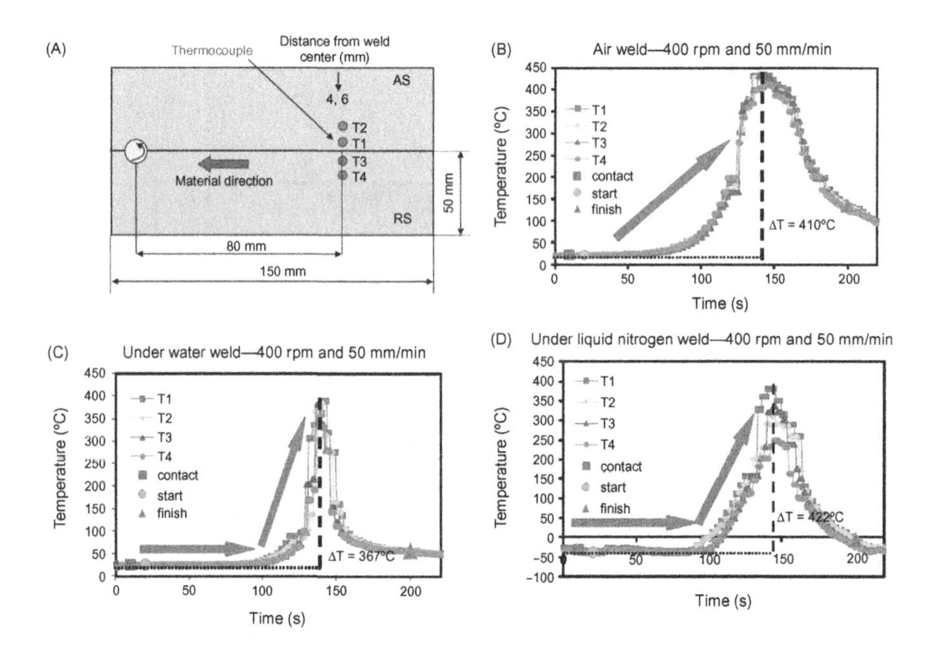

Figure 5.14 Temperature measurement and profiles of dissimilar butt welds between AA5083 and AZ31, (A) schematic diagram showing the location of thermocouples and temperature profiles of welds made, (B) in air, (C) in water, and (D) under liquid nitrogen (Mofid et al., 2012, Reprinted with permission from Springer).

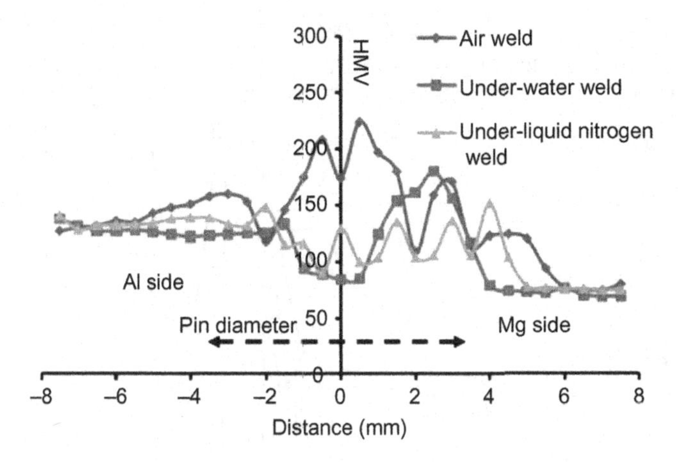

Figure 5.15 Mircohardness profiles for welds shown in Figure 5.14 (Mofid et al., 2012, Reprinted with permission from Springer).

5.2 Al TO Cu

Dissimilar welding between Al and Cu has also received significant attention recently, mainly because of the potential application in electronic components and power generation industries by taking advantage of the lightweight and low-cost aluminum alloys and the superior electric and thermal conductivity of copper. One example of potential application of dissimilar Al and Cu welds is the electrical connections, such as battery assembly. Similar to dissimilar welding between Al and Mg or steel, high metallurgical reactivity and affinity between Al and Cu leads to the formation of hard and brittle IMCs as shown in Al–Cu phase diagram (Figure 5.16) at the interface, which in turn, results in low weld strength. It is quite challenging to achieve defect-free friction stir welds between Al alloy and Cu. Large void, cracks and other distinct defects are frequently present in the welds. Research activities mainly involve process development for optimizing the weldability, and mechanical as well as electronic performance of dissimilar welds between Al alloys and pure copper or copper alloys. Particularly, placement of workpiece on advancing or retreating side, with or without tool offset, direction of tool offset (either to Al or Cu), welding parameters (rotational rate and welding speed) are the key factors that have been studied since they affect the type of IMC produced as well as the distribution of IMC in the welds.

Similar to what has been observed in dissimilar Al and Mg FSW, the placement of workpiece plays a more important role in achieving

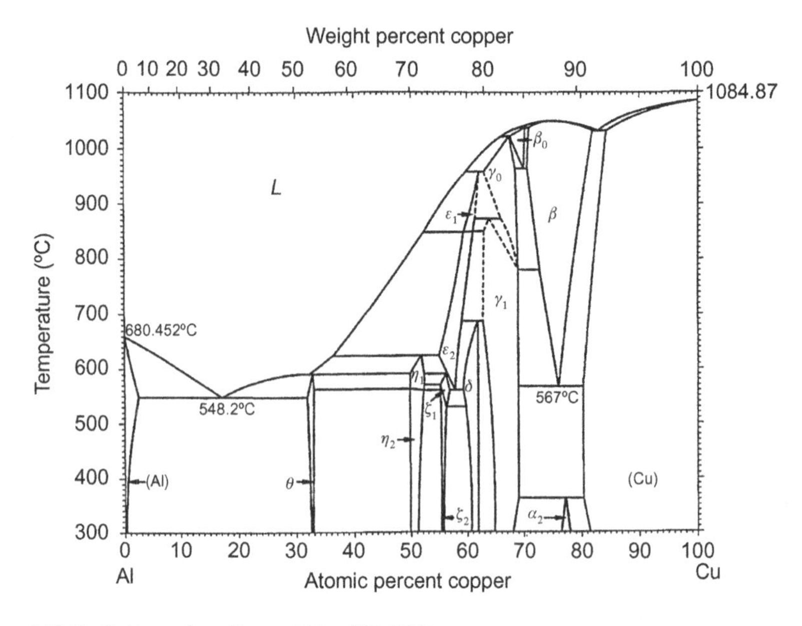

Figure 5.16 Al–Cu binary phase diagram (Massalski, 1990).

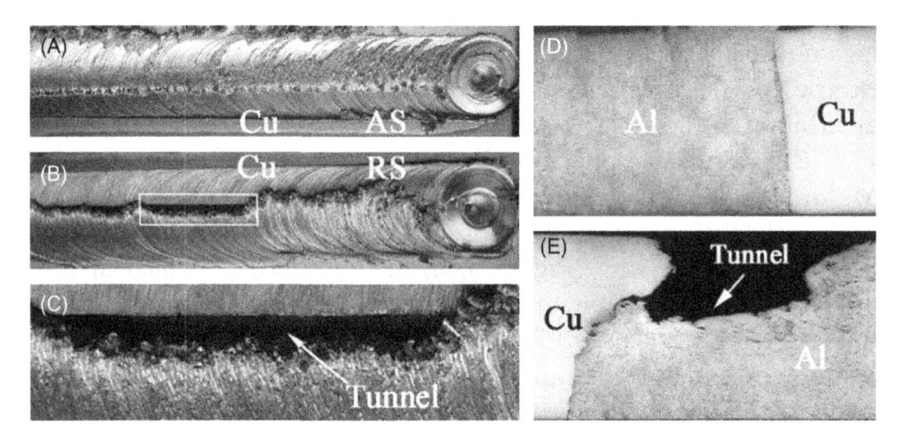

Figure 5.17 Surface morphologies and cross sections of dissimilar butt-welded AA1060 and Cu with Cu plate fixed at (A) advancing side, (B) retreating side, (C) magnified view of tunnel defects in the weld having Cu on the retreating side, (D) cross section for the weld shown in (A), and (E) cross section for the weld shown in (B) (Xue et al., 2011).

sound welds between Al and Cu. Figure 5.17 shows friction stir butt-welded 5 mm AA1060 and commercially pure Cu with either Al or Cu on the advancing side. Both welds were produced at 800 rpm and 100 mm/min tool traverse speed using a tool with 20 mm diameter shoulder and 6 mm diameter pin. Tool pin offset of 3 mm to the Al side was used in both cases. It can be observed that a good defect-free weld is possible only when the harder Cu plate was fixed on the advancing side. On the other hand, a large volume defect is visible when the softer Al plate was placed on the advancing side. Similar findings have also been reported in other studies especially for softer and harder material systems, such as Al to Cu, Al to steel. Different behaviors are likely related to the ease of flow of the softer material (Al) when it was placed on the retreating side so that it can be transported to the advancing side, in turn resulting in good mixing of dissimilar materials. When harder Cu was placed on the retreating side, it can cause softer Al to be extruded from the weld zone due to the difficulty of flow of harder Cu toward the advance side and lead to large volume defects in the welds.

The pin offset affects the material mixing and IMC formation in dissimilar Al and Cu welds. Sound dissimilar welds could be produced under a larger pin offset to the softer material. Figure 5.18 shows the surface morphologies of butt-welded AA1060 and Cu under various tool offset into Al. A poor surface finish is obvious without tool offset or a small offset of 1 mm. Increasing offset improves the surface quality. Cross-sectional view indicates voids and large Cu debris at a small offset. The large Cu debris is hard to deform, flow, and mix with the Al. This led to the poor surface quality, bonding, and formation of voids as well as IMCs. With larger tool offset, only Cu pieces with

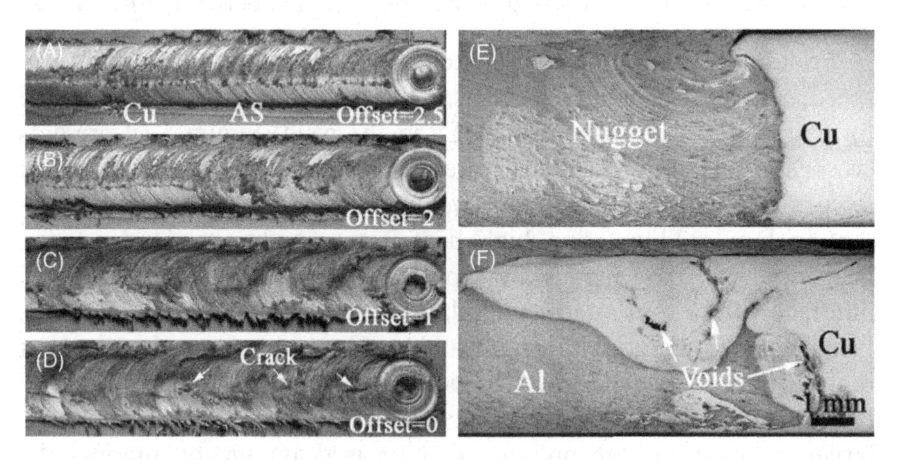

Figure 5.18 Surface morphologies of butt-welded AA1060 and Cu under various tool offset into Al: (A) 2.5 mm, (B) 2 mm, (C) 1 mm, and (D) 0 mm. Cross sections for welds with 2.5 and 1 mm offset are shown in (E) and (F), respectively (Xue et al., 2011).

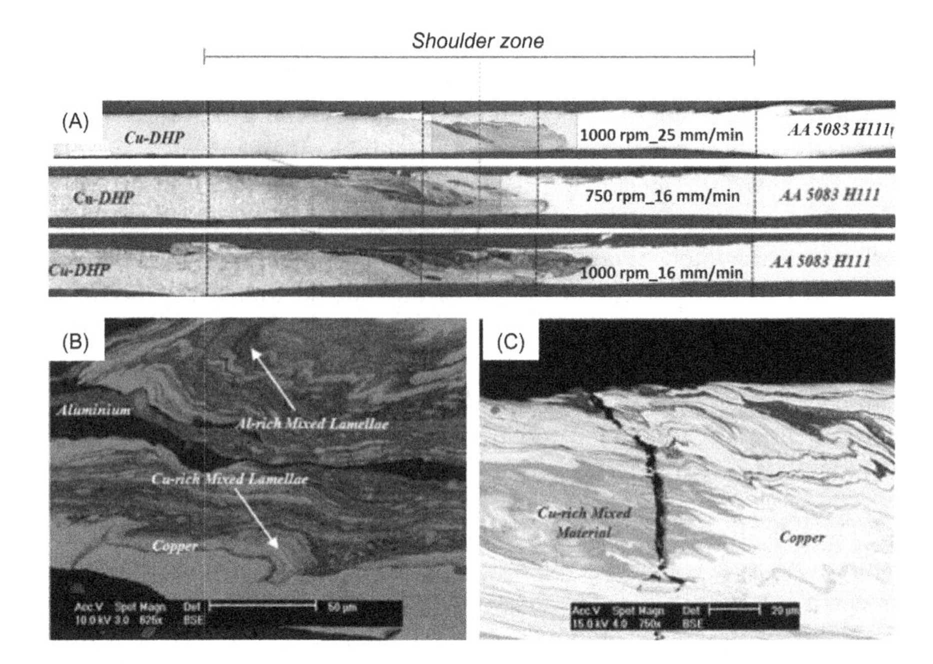

Figure 5.19 Friction stir butt-welded AA5083 to oxygen-free Cu: (A) cross sections under three welding parameters combination, (B) SEM image of weld zone made at 750 rpm and 16 mm/min, and (C) SEM image of weld zone produced at 1000 rpm and 16 mm/min (Galvão et al., 2011, Reprinted with permission from Maney Publishing).

relatively small size were scratched from the Cu. In this case, it is easier for the small Cu pieces to mix and react with the Al in the stir zone, thus, sound metallurgical bonding can be obtained between Al and Cu.

Frictional heat input dominated by tool rotation rate, traverse speed, and tool geometry influences the weld integrity as well as the formation and distribution of IMCs in the weld. Figure 5.19 presents cross sections and microstructures of butt-welded AA5083 to an oxygen-free Cu. Three different welding parameters were used representing low, medium, and high heat input conditions. Under lower heat input, the base metals are separated by a sharp and well-defined interface with a thin intermetallic layer distribution along Al/Cu interface. A mixed region with bright and dark zones is noticed when heat input is increased. The mixed region shows complex mixing between lamella of Al and Cu. Further increase in the heat input promoted a larger mixed zone with presence of increased amount of intermetallic rich structure and a crack propagating along the mixed patterns.

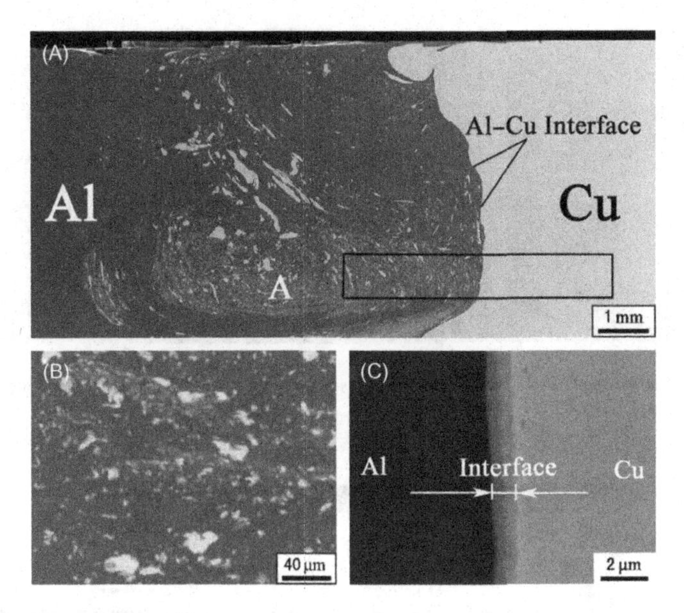

Figure 5.20 SEM observation of friction stir butt-welded AA1060 to pure Cu: (A) cross-sectional view of the weld, (B) magnified view of the Cu particle rich region, and (C) microstructure at the interface between Al and Cu (Xue et al., 2010).

IMC structures have a significant influence on the mechanical properties of welds. A thin IMC layer or structure could improve the mechanical properties. Figure 5.20 shows friction stir butt-welded 5 mm AA1060 to pure copper. The welding was conducted with Cu plate fixed on the advancing side and tool pin offset into the Al side. A tool rotation rate of 600 rpm and a tool traverse speed of 100 mm/min were used. Figure 5.20A shows the cross-sectional view of the welds exhibiting a mixture of irregular Cu particles inhomogeneously distributed in the Al matrix. A Cu particle-rich zone composed of IMC is observed at the bottom of the stir zone. In addition, a continuous and uniform discernible sublayer of about 1 μm thick can be noticed at the interface between Al and Cu. The thin layer was later identified as Al_2Cu and Al_4Cu_9 and proved to be stronger than the heat-affected zone, which yield a high tensile strength of about 80% of the base Al alloy. Figure 5.21 presents the bending behavior of butt-welded Al–Cu shown in Figure 5.20. The dissimilar welds could be bent to 180° without sign of fracture based on SEM observation. More importantly, no crack was observed in the IMC layer as shown in Figure 5.21B.

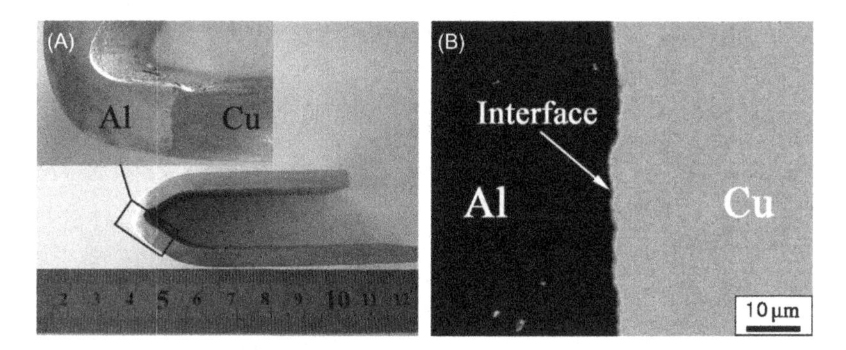

Figure 5.21 (A) Friction stir butt-welded AA1060 to pure Cu after bending test and (B) microstructure of the interface (Xue et al., 2010).

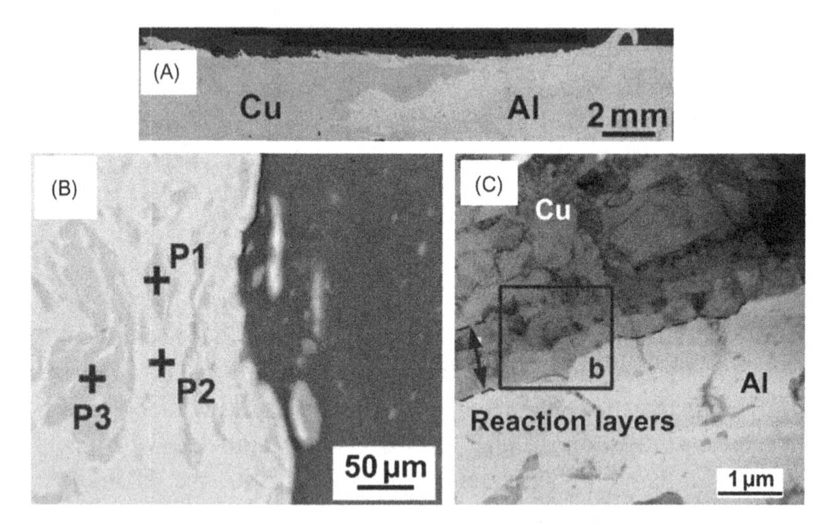

Figure 5.22 Friction stir butt-welded Al alloy 5A02 to pure copper: (A) transverse cross-sectional view, (B) SEM image for the Al and Cu interface, and (C) TEM observation of the interface showing thin reaction layers (Tan et al., 2013).

Another example considered here is regarding a butt welding between 3 mm Al alloy 5A02 and pure copper. Different from previous examples, the welding was conducted with Al on the advancing side and a tool offset of 0.2 mm to the Al alloy side. A tool rotation rate of 1100 rpm and a traverse speed of 20 mm/min were used to generate defect-free welds. It is noteworthy that a higher tool traverse speed does induce channel defect. Figure 5.22A shows the macrostructure of the weld with some material mixing. High-magnification view of Al and Cu interface indicates obvious contrasts, which were later

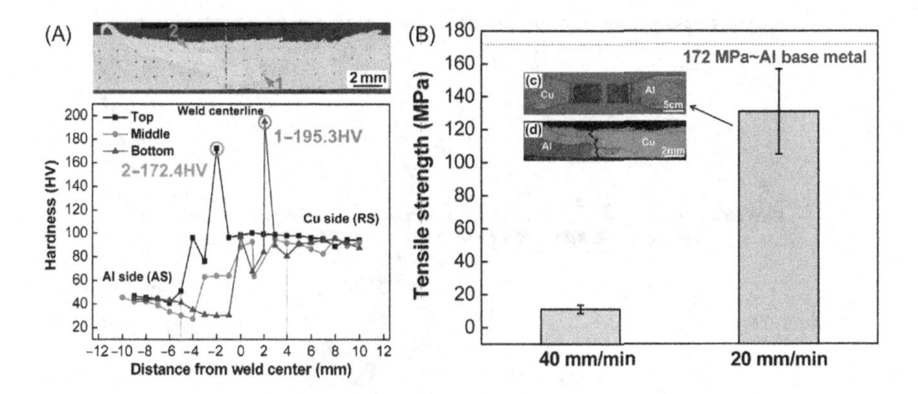

Figure 5.23 (A) Hardness profile for the weld produced at 20 mm/min shown in Figure 5.21 and (B) tensile strength of welds made at 20 and 40 mm/min (Tan et al., 2013).

identified as formation of IMCs and solid solution. A thin and continuous layer with thickness of about 1 μm with fine grains was clearly observed in the TEM micrograph (Figure 5.22C).

Figure 5.23 shows the hardness profiles and tensile strength of welds presented in Figure 5.22. An inhomogeneous hardness distribution in the weld region can be observed in Figure 5.23A. The high hardness values above 170 HV were attributed to the formation of nanostructured IMCs. Figure 5.23B presents the tensile strength of welds made at 20 and 40 mm/min and compared with that of base Al alloy. A high tensile strength of 130 MPa, corresponding to 75.6% of the Al alloy base metal, was achieved with the fracture located along the Al and Cu border and extended into the stir zone.

5.3 Al TO STEEL

As discussed in Chapter 2, welding between Al alloys and steels fall into the third category where metallic materials being joined are not only different from each other in terms of base materials but also have very different thermo-physical properties such as melting points. In such scenario, the welding between such disparate combinations of materials become very difficult and requires special joining strategies.

Unlike welding of dissimilar alloys with similar base materials or even those which have very similar melting points (or range), for Al alloy and steel combination the welding tool cannot be plunged symmetrically into the joint line. It will result in excessive heating of steel

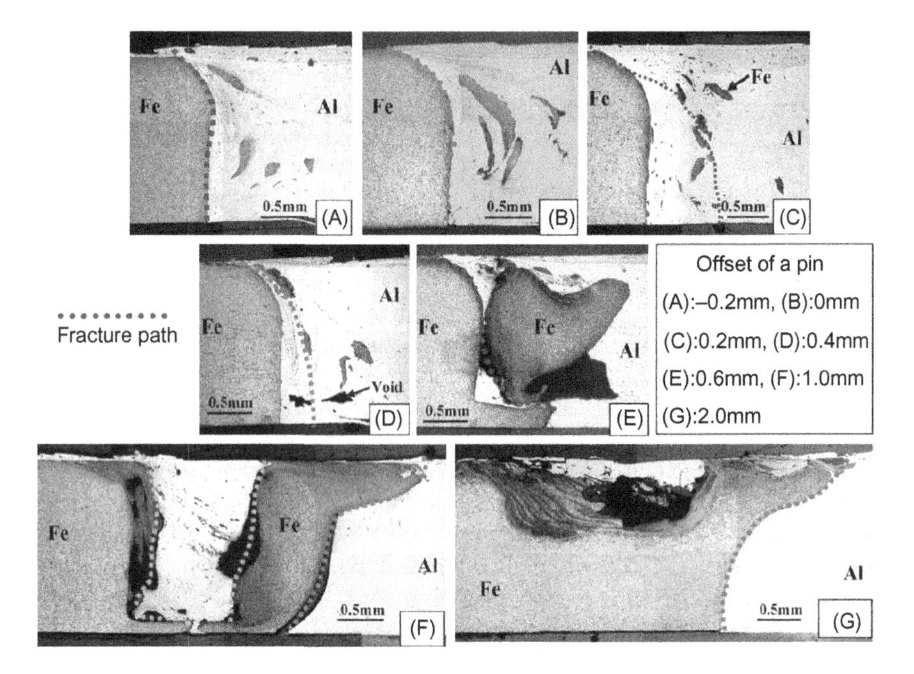

Figure 5.24 Effect of tool pin offset on the joint integrity (Watanabe et al., 2006).

which will cause melting of the Al alloy. This, of course, will result in a defective weld. Hence, for joining Al alloys and steels, the offset method is used during FSW. In offset method, the rotation center of the tool is biased toward the low-melting-temperature material, and in some cases to the extent that it almost plunges inside the Al alloy. The work done by Watanabe et al. (2006) is shown in Figure 5.24. It shows the effect of tool pin offset on the joint integrity. The tool offset varied from the cylindrical surface of the tool pin being 0.2 mm away from the faying surface toward Al alloy to 2 mm inside steel workpiece. Among all, the weld with 0.2 mm pin offset toward steel exhibited maximum tensile strength. The tensile test results for all the welds based on this study are shown in Figure 5.25. Clearly, the joint efficiency of the weld corresponding to 0.2 mm offset toward steel side is approximately 85% (of the Al alloy).

When tool plunges completely inside Al alloys, such welds can be made using the tool used for making joining low-melting-point materials such as Al and Mg alloys. But, when offset is such that tool is making contact with the high-melting-temperature material, it becomes

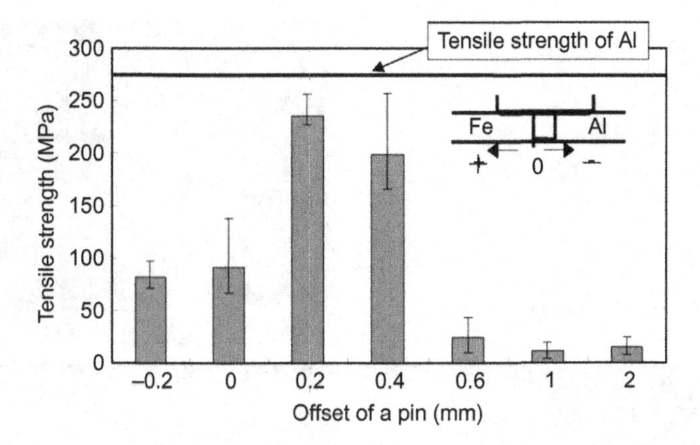

Figure 5.25 Tensile strength versus tool pin offset for welds shown in Figure 5.24 (Watanabe et al., 2006).

important to use ceramic-based tools or tools made of refractory metals.

Figure 5.26 shows dissimilar weld between HC260LA high-strength steel and AA6181-T4 Al alloy. It also includes elemental mapping across the welded region using energy dispersive x-ray spectroscopy (EDS). Figure 5.26A shows a zigzag but sharp interface between these two alloys, indicative of no mixing. Al, Fe, and Zn elemental mapping (in Figure 5.26B–D) also confirm this observation. Zn elemental mapping also shows a vortex- like feature showing diffusion of Zn in Al alloy (left portion of the figure which is the retreating side). In this welding, pin surface was about 1 mm away from the Al–steel joint line. Although in this configuration tool never comes in contact with the steel, a W-Re25 tool was used for processing this combination of materials. The microhardness measurement carried out for this weld across the welded zones is shown in Figure 5.27. It shows a sharp discontinuity at the weld interface. Tensile testing of the weld showed a strength level slightly smaller than AA6181-T4 alloy but ductility value significantly smaller than Al alloy.

The effect of intermetallic formation on the joint strength in Al alloy and steel weld has been evaluated by Tanaka et al. (2009). Figure 5.28 shows tensile strength versus IMC thickness plot. It is evident from this plot that tensile strength of the joint increased with a decrease in IMC thickness. It, therefore, suggests that every care should be taken to minimize the formation of IMC. For example, Tanaka et al. (2009) also investigated the dependence of tensile strength of the dissimilar Al alloy

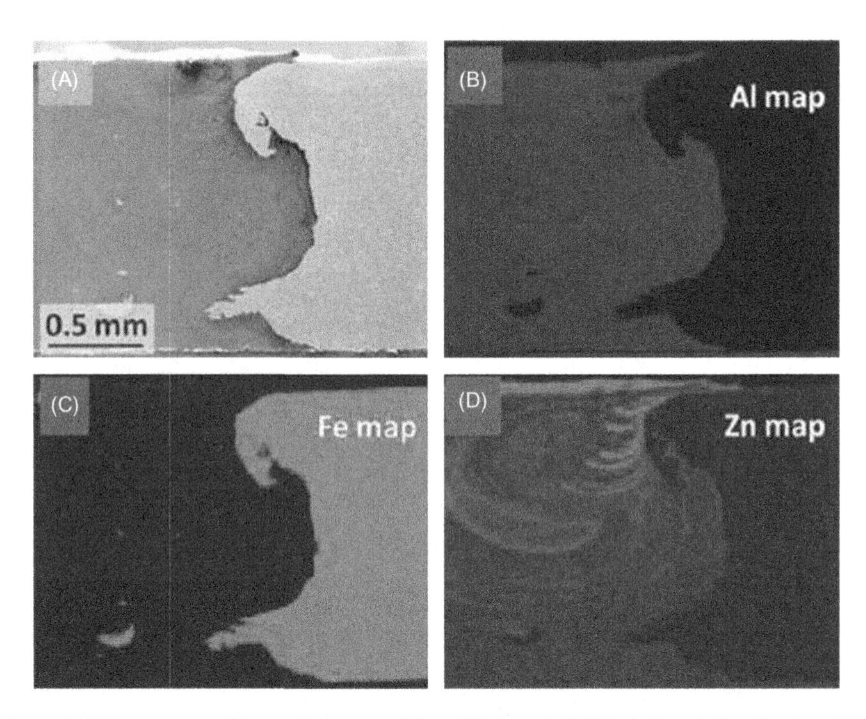

Figure 5.26 Cross section and elemental mappings of the weld between HC260LA high-strength steel and AA6181 (Coelho et al., 2012).

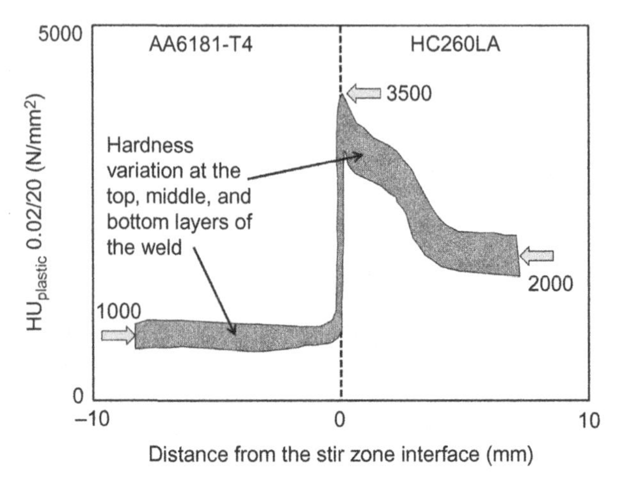

Figure 5.27 Microhardness profiles of the weld between HC260LA high-strength steel and AA6181 (Coelho et al., 2012).

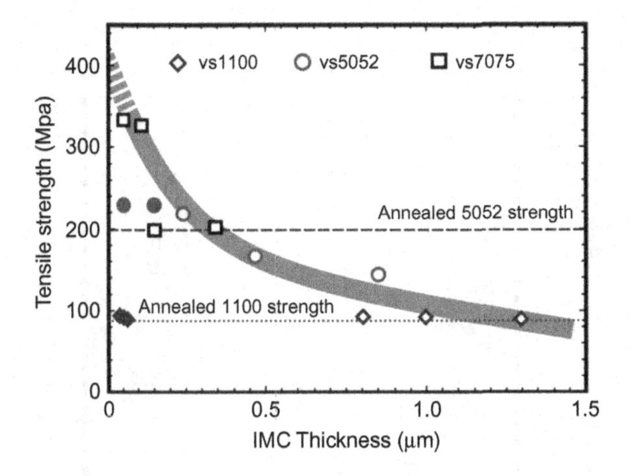

Figure 5.28 Effect of IMC thickness on tensile strength of the friction stir weld between aluminum alloys and steel (Tanaka et al., 2009).

and steel joints as a function of heat input, and it was found that the tensile strength increased as heat input decreased. This finding indicated that IMC layer thickness was proportionally related to heat input. Hence, in order to decrease the thickness of IMC layer, the FSW runs should be carried out at low heat input, that is, at low tool rotation rate (ω), high tool traverse speed (ν), or high ν/ω. The work carried out by Liu et al. (2014) also supports the observation made here. They have shown that for a given tool traverse speed, the IMC thickness increased with increase in tool rotation rate, and for a given tool rotation rate, the IMC layer thickness decreased with increase in tool traverse speed.

The mechanism of joint formation between Al and steel was proposed by Watanabe et al. (2006). It is schematically shown in Figure 5.29. They discussed that the sound joint formation is not possible if steel is placed on the retreating side (Figure 5.29B) because the plasticized Al alloy makes contact with the oxide film on the faying surface of steel on the retreating side. Moreover, the tool tangential velocity on the retreating side will try to peel off already deposited layer on the retreating side. Hence, a sound weld formation is very difficult. When steel is on advancing side, the tool makes contact with the steel and disrupts the continuous oxide film thereby exposing an activated layer of Fe to Al. The plasticized Al alloy makes contact with this freshly exposed Fe atoms and gets pressed by the rotating tool toward the steel side. It results in formation of good joint between both materials.

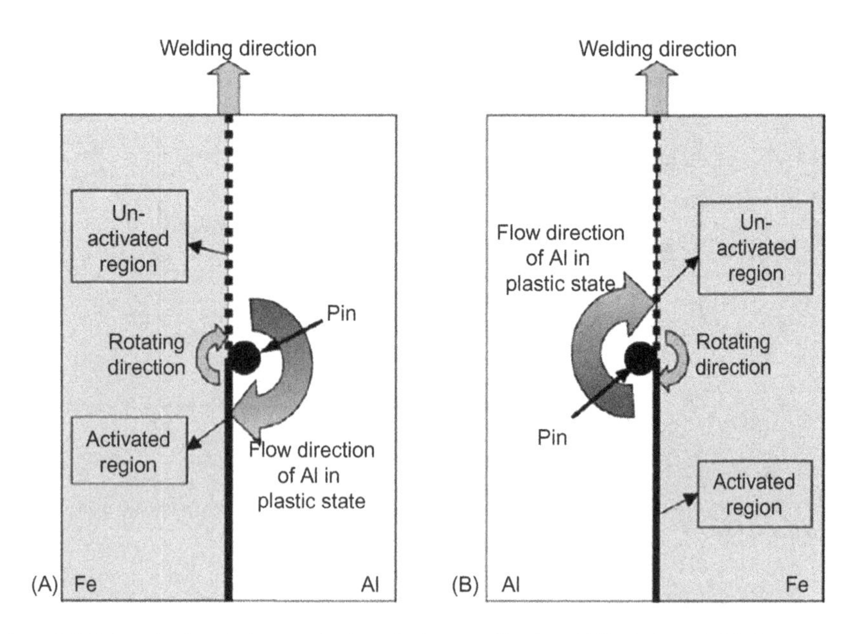

Figure 5.29 Schematic of mechanism on joint formation between Al and steel (Watanabe et al., 2006).

In Chapter 2, we have discussed that if the faying surface is present on retreating side, due to low level of deformation, it is difficult to disrupt the oxide at the faying surface. Hence, an adequate bonding does not take place between plasticized Al and the oxide film, which is an oxide layer. Other factor which influences the bonding between the Al alloy and steel is the softening of steel due to its presence on the retreating side. Since the steel is present on the retreating side, temperature rise is low as compared to Al alloy and that also makes plastic flow of steel very difficult and leads to no bonding between the alloys.

All the above mentioned weld characterizations are limited to static testing. Uzun et al. (2005) have studied the fatigue property of dissimilar friction stir welds between AA6013 and X5CrNi18-10 stainless steel. The result of the dissimilar weld was compared with the S–N curve for AA6013-T6 alloy. Overall, the joint of the dissimilar weld had similar fatigue property to the base material AA6013.

5.4 Al TO Ti

The welding of dissimilar Al to Ti is also of interest to transportation industries for integrated lightweight, but strong and corrosion-resistant

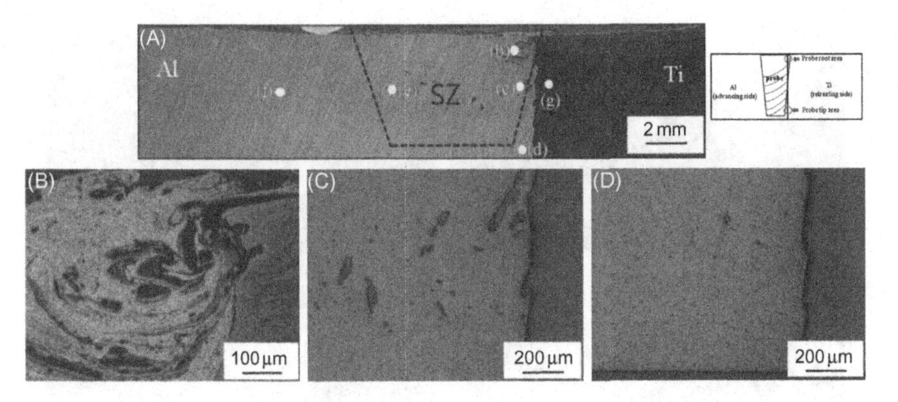

Figure 5.30 Friction stir butt-welded Al alloy 6061 and Ti–6% Al–4% V with (A) transverse cross section and (B–D) showing local microstructure near the butt interface (Bang et al., 2011, Reprinted with permission from The Japan Institute of Light Metals).

components or structures. As specified in Chapter 2, significantly different melting temperatures, physical properties, and high chemical affinity between Al and Ti alloys at elevated temperatures complicate such dissimilar welding to achieve sound weld integrity and high strength. Like dissimilar Al to steel FSW, tool offset away from the interface is always adopted in such welding process to reduce tool wear and avoid overheating of softer Al alloys; however, tool pin is required to have direct contact with and scribe the harder Ti to form necessary material mixing and metallurgical bonding. This places critical requirement for the tool material to be wear-resistant. Hybrid welding process combining FSW with additional heating methods such as arc welding has also been employed to heat Ti in order to reduce tool wear and improve mixing. Al and Ti IMCs, such as $TiAl_3$, have also been observed at the interface and are believed to influence the mechanical performance.

Figure 5.30 shows cross section and microstructure of dissimilar-welded 5 mm thick AA6061 to a same thickness Ti alloy Ti–6Al–4V in a butt configuration. In this particular case, the softer Al alloy was placed on the advancing side and the tool made of standard tool steel offset to the Al side in order to keep slight contact between the pin and the Ti alloy. The tool used has a 14 mm diameter shoulder and a 4.5 mm long tapered pin. A tool rotation rate of 1000 rpm and tool traverse speed of 150 mm/min were employed. The cross section and microstructure indicate some material mixing between Al and Ti near the interface without welding defects. Since the pin is mainly placed

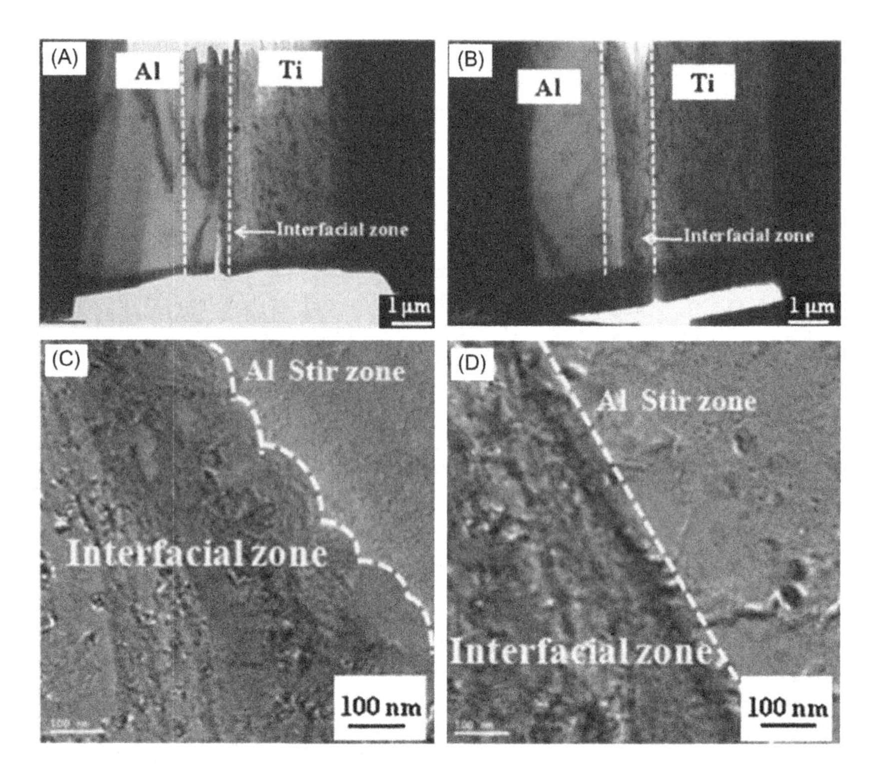

Figure 5.31 TEM observation of the interface near (A and C) tool root region and (B and D) tool tip region. Interfacial layer zone can be noticed in both the regions (Bang et al., 2011, Reprinted with permission from The Japan Institute of Light Metals).

inside of the Al, only small pieces of Ti were scribed out of the bulk Ti. For the bottom region, where pin does not have direct contact with Ti, no mixing but a clear boundary can be observed.

Figure 5.31 illustrates the TEM observation at the weld interface near the pin root and the pin tip regions. It can be noted that a thin interfacial layer of about 1 μm thickness is formed in both regions. Wavy interface is noticed for the interface near the root but not for the tip. This is suggested to be related to the shape and surface features of the tool pin. The chemical identification of the interfacial zone was not reported in this study.

The tool offset with respect to the butt interface affects interfacial microstructure and mechanical properties. One example shown here is the butt welding of 2 mm AA6061 and Ti alloy Ti–6Al–4V with the hard Ti alloy on the advancing side. Figure 5.32 shows the cross-sectional views of weld interfaces under various tool offsets from 0.3 to 1.2 mm.

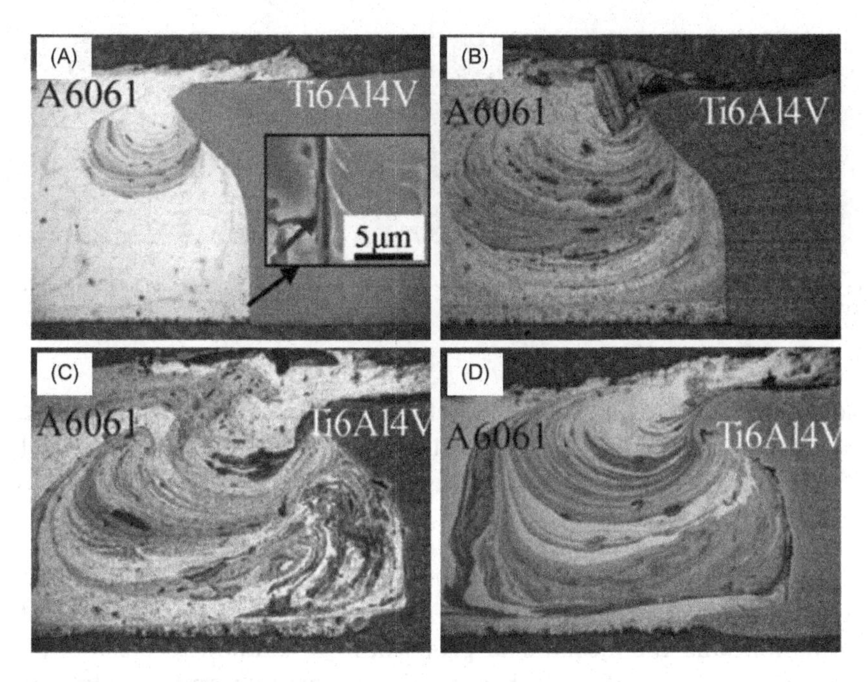

Figure 5.32 Cross sections of butt-welded AA6061 to Ti6Al4V with various tool offsets: (A) 0.3 mm, (B) 0.6 mm, (C) 0.9 mm, and (D) 1.2 mm (Song et al., 2014).

It is noteworthy that the tool pin mainly plunges into the softer Al side and the offset indicates the distance pin edge into Ti alloy. All welds were produced under 1000 rpm and 120 mm/min by using a WC-Co tool with 15 mm shoulder diameter, 6 mm pin diameter, and 1.9 mm pin length. All cross sections show swirl-like structure. The area of this swirl-like structure increases with the tool offset distance. The analyzed local chemical compositions indicate distribution of Ti particles in an Al matrix. The fraction of Ti particles appears to increase with tool offset. It is obvious that the tool offset distance exhibits a great influence on the material mixing near interface. When the offset is less than 0.6 mm, the bottom of the interface is not bonded.

Figure 5.33 shows high-magnification views of interface at mid-thickness of the welds with 0.6, 0.9, and 1.2 mm tool offsets. An IMC layer with thickness of about 0.5 μm is suggested for 0.6 and 0.9 mm offsets based on the image contrast. The IMC layer becomes thicker and complex at 1.2 mm tool offset. Alternative band structure with IMC, Ti, and micro-crack is noticed near the interface on the Ti alloy

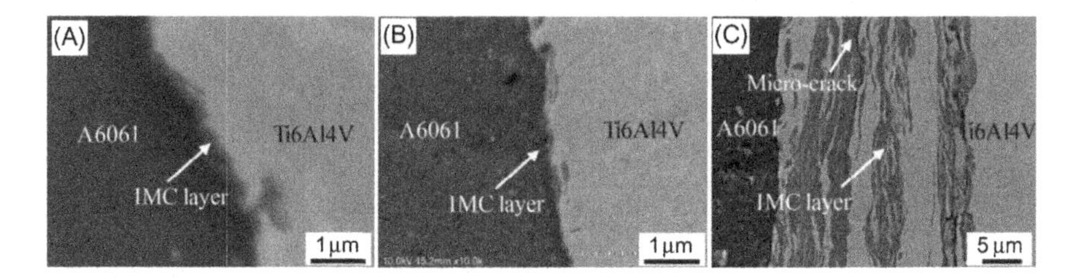

Figure 5.33 SEM images of the weld interface at middle thickness for welds with tool offset of (A) 0.6 mm, (B) 0.9 mm, and (C) 1.2 mm (Song et al., 2014).

side. The increased amount of IMC and the presence of micro-crack at 1.2 mm tool offset leads to a strength reduction of approximately 35 MPa compared to the average tensile strength of 197 MPa achieved with 0.9 mm tool offset.

As shown above, the proper amount of material mixing between Al and Ti plays an important role in the interface morphology, IMC formation, and the mechanical properties. In addition to the tool offset, welding parameters such as tool rotation rate and tool traverse speed also affect the material mixing. An example shown here is to evaluate the tool traverse speed on butt welding high-strength Al alloys AA2024 and AA7075 to pure Ti. A tool made of tool steel was mainly plunged into Al. All welds were made using 850 rpm tool rotation rate. Figure 5.34 shows microstructures of weld interface between Al alloys and Ti. Blank regions containing a mixture of Al alloy, pure Ti, and IMC ($TiAl_3$) are observed at the weld interface. The details about IMC distribution were not reported. It can be observed that compared to the lower traverse speed, the width of mixed region is wider at higher tool traverse speed for both Al alloys. It is noteworthy that this mixed region occurs only partly at the interface, especially the upper half of the plate. Similar to what has been shown in Figure 5.32 where mixing of Al alloy and Ti starts from the upper portion of the welds. Figure 5.35 shows the tensile properties of dissimilar welds between Al2024/Al7075 and Ti under three traverse speeds. High tensile strength can be observed for welds with high traverse speed, which appears to be related to the material mixing and IMC formation. Relatively low tensile strength for Al 7075/Ti welds compared to Al 2024/Ti welds is attributed to insufficient material flow and material mixing.

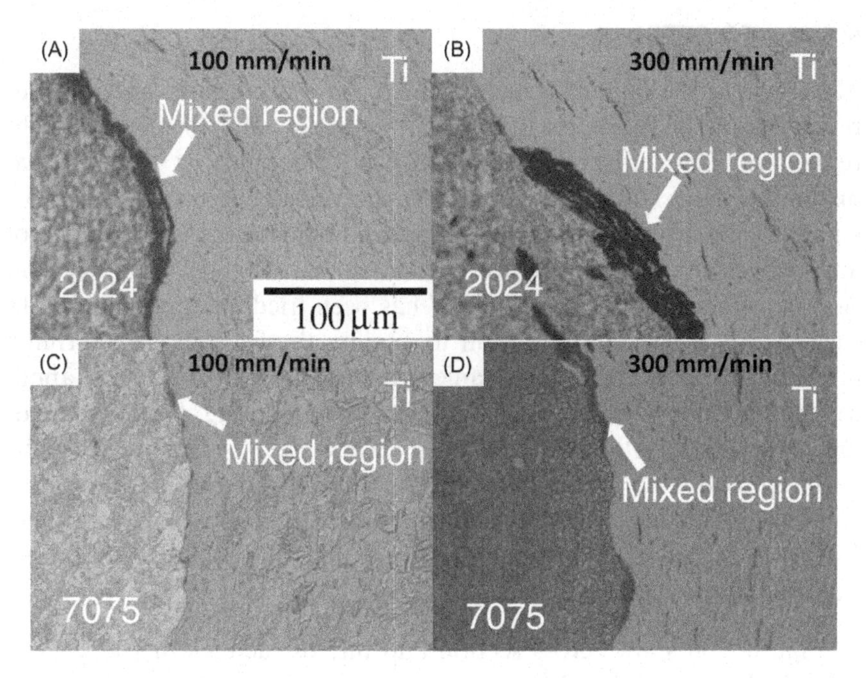

Figure 5.34 Microstructure of butt-welded interface between (A and B) Al alloy AA2024 and pure Ti, and (C and D) Al alloy AA7075 and pure Ti (Aonuma and Nakata, 2011, Reprinted with permission from The Japan Institute of Light Metals).

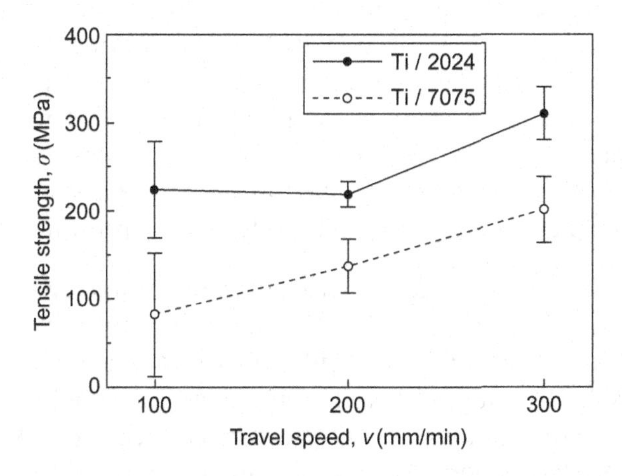

Figure 5.35 Tensile strength as a function of tool traverse speed for butt welds shown in Figure 5.34 (Aonuma and Nakata, 2011, Reprinted with permission from The Japan Institute of Light Metals).

5.5 Mg TO STEEL

Mg alloys as the lightest structural metals are promising candidates to replace some of the steels in automotive application to realize weight reduction. This brings in the challenge of joining or welding two dissimilar materials with quite different melting temperatures, flow strengths, immiscibility in both solid and liquid states, and absence of any congruent melting phases. So strong bonding between Mg and steel is problematic. Recently, FSW has been tried on dissimilar Mg to steel mainly in lap configuration to explore the feasibility on forming metallurgical bonding or mechanical interlocking for structural application. IMC between Mg and steel is not likely during dissimilar welding, formation of other IMC between Al and steel, or Mg and Zn have been reported which depends on the Al concentration in Mg alloy, surface coating of steel, and the welding conditions. The influence of these IMCs on mechanical properties of dissimilar Mg and steel has been explored and discussed.

Figure 5.36 shows cross section and microstructures of friction stir spot-welded Mg alloy AM60 to dual phase steel DP600. The thickness of AM60 was 1.2 mm and 1.8 mm for DP600. The DP600 was coated with a Zn layer. Mg alloy was placed on top of steel during welding with a tool rotation rate of 3000 rpm and a plunge speed of 1 mm/s. A tool made of W-25Re alloy was employed. It had a shoulder diameter of 10 mm, a pin diameter of 4 mm, and a pin length of 1.7 mm. Cross-sectional view indicates upward flow of the lower steel into the upper Mg alloy and formation of bonding or interlocking between them. A clear boundary can be seen due to the contrast. High-magnification views of the boundary show a sharp transition between the Mg and steel without any sign of IMCs. However, a cast structure was observed which was further identified as an Mg–Zn eutectic structure. The eutectic reaction occurs at a relatively low temperature of 339°C.

The effect of Zn coating on weld interface characteristics has been studied by Schneider et al. (2011) by friction stir lap welding 2 mm Mg alloy AZ31 to 2 mm mild steel with and without zinc coating. A tool made of W-25Re alloy with a 12.7 mm scrolled shoulder and 2 mm long square shaped pin was used for dissimilar welding of Mg and steel with both surface conditions. Welds made at a tool rotation rate of 1200 rpm and a traverse speed of 500 mm/min are shown in Figures 5.37 and 5.38 for comparison. Figure 5.37 shows the weld

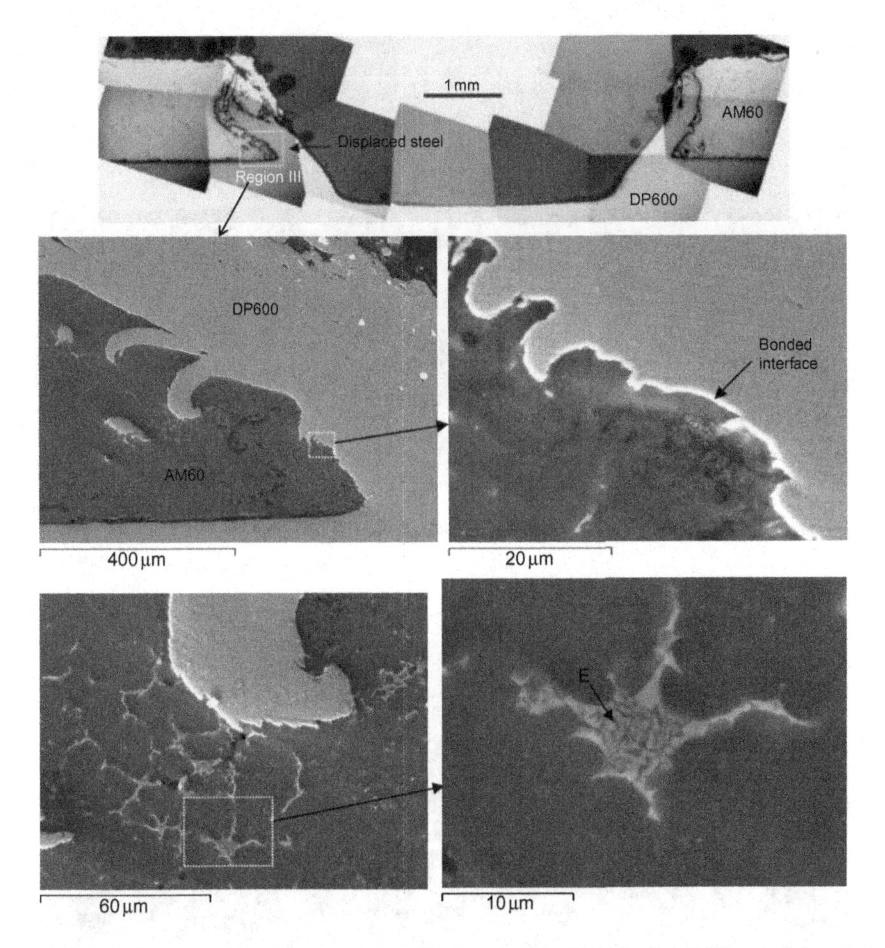

Figure 5.36 Cross section and local microstructures of friction stir spot-welded Mg alloy AM60 to dual phase steel DP600 made at 3000 rpm, 1 mm/s plunge speed, and 0.55 mm pin penetration into bottom steel (Liyanage et al., 2009, Reprinted with permission from Maney Publishing).

between AZ31 and steel without Zn coating, and Figure 5.38 shows the weld with coating. In both cases, the Mg alloy was placed on the top of the steel, and no defects were detected for either of them. For the weld without Zn coating on the steel, many small size steel particles can be noticed near the interface. The composition analysis does not show any indication of IMC at the interface, but high content of iron in the Mg sheet. For the weld with Zn coating on the steel, large irregular shaped steel particles were identified near the interface. In addition, a layer of about 50 μm thickness is present at the interface, more likely on the Mg side. Within the layer, a eutectic Mg–Zn structure was detected as shown in Figure 5.38C.

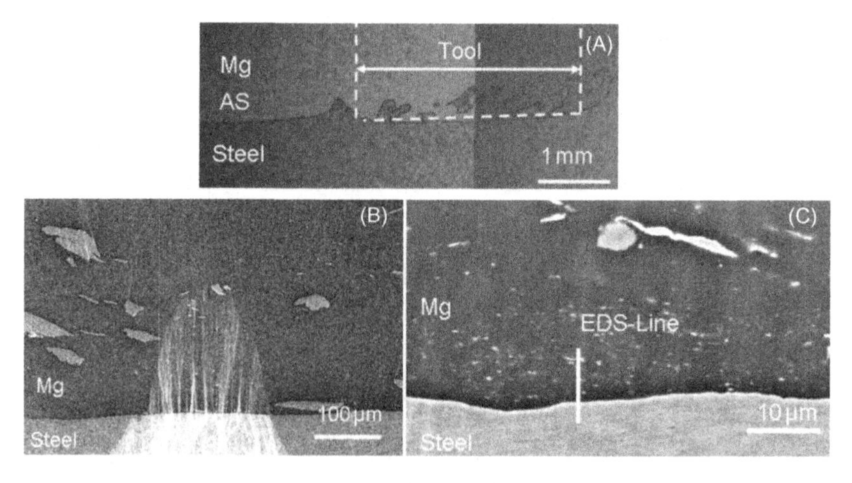

Figure 5.37 Cross section and microstructures of friction stir lap-welded Mg alloy AZ31 to mild steel without surface coating (Schneider et al., 2011, Reprinted with permission from Maney Publishing).

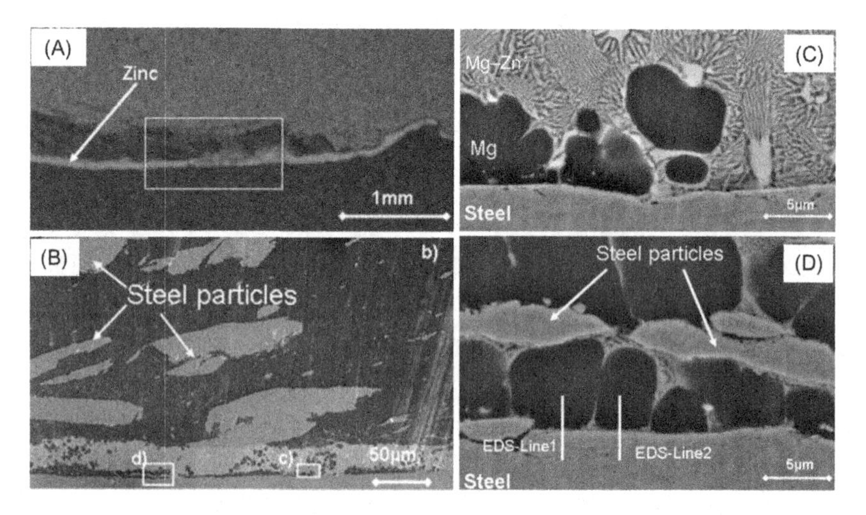

Figure 5.38 Cross section and microstructures of friction stir lap welded Mg alloy AZ31 to Zn-coated mild steel (Schneider et al., 2011, Reprinted with permission from Maney Publishing).

Feasibility study on lap welding of Mg alloy to high-strength low-alloy (HSLA) steel has been conducted by Jana et al. (2010). Mg alloy AZ31 of 2.33 mm thickness and galvanized HSLA steel of 1.5 mm thickness were lap welded using a tool made of H13 tool steel. The tool has a scrolled convex shoulder with diameter of 12.8 mm and a 2.34 mm long pin with diameter of 4.9 mm. The tool tip was

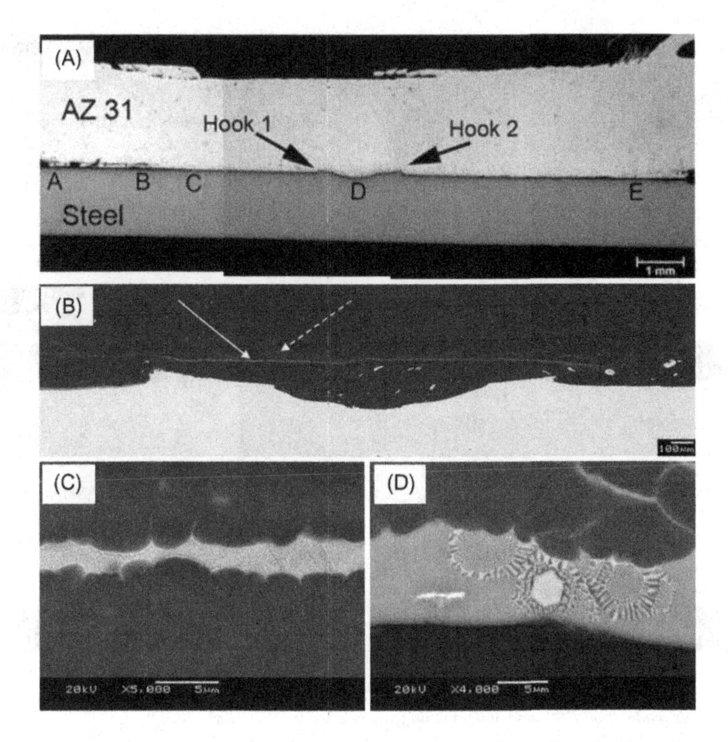

Figure 5.39 Cross section and microstructures of friction stir lap-welded Mg alloy AZ31 to galvanized HSLA steel (Jana et al., 2010, Reprinted with permission from Springer).

programed to plunge 2.45 mm. Figure 5.39 shows cross section and local microstructures of the dissimilar weld made at 700 rpm and 100 mm/min tool traverse speed. Hooking features are visible due to the plunge of pin into HSLA steel. Figure 5.39B–D shows SEM images captured in backscattered composition mode. No Zn layer or any new phase along the Mg/steel inference can be observed at the weld centerline from Figure 5.39B. The stir zone above interface shows presence of small steel particles due to the scribe impact of pin. A very bright thin layer above hooks can be observed which is identified to be Zn-rich. High-magnification views indicate lamellar structure inside. The weld under tensile shear testing indicates an interfacial failure with an average failure load of 6.3 kN based on 30 mm wide specimen.

The tool design also plays an important role on interface morphology and affects microstructural evolution. In another work conducted by Jana and Hovanski (2012), same dissimilar Mg/steel as shown in Figure 5.39 was evaluated by using another tool design. The new tool had similar convex scrolled shoulder, but a stepped spiral pin with a

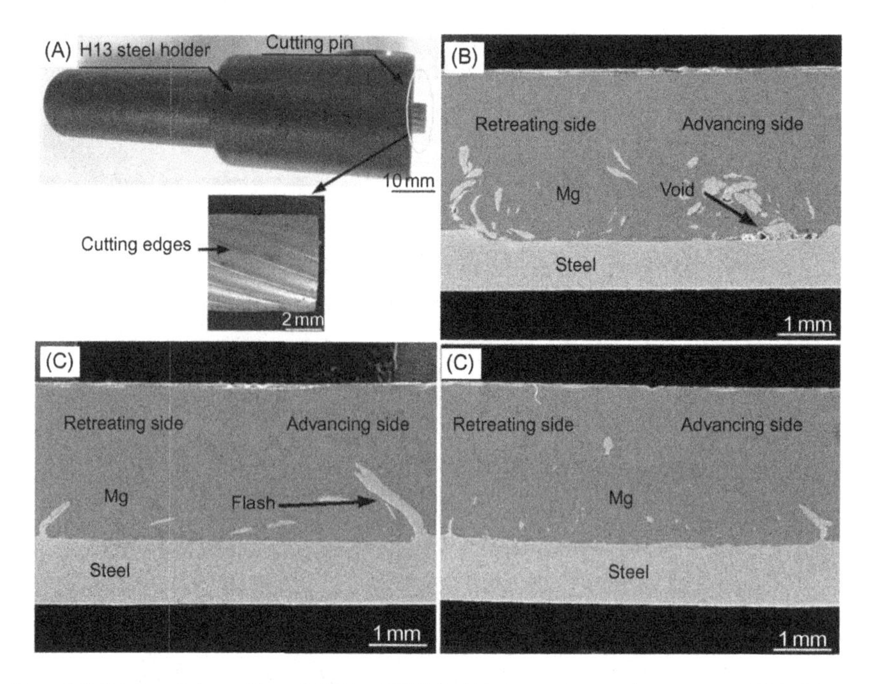

Figure 5.40 Friction stir lap welding of AZ31 to SUS302: (A) picture of tool and cutting pin and (B–D) cross sections of welds made at 60, 190, and 300 mm/min traverse speeds (Wei et al., 2012).

short WC insert at the pin bottom. This WC insert was employed to scribe the lower steel sheet. Welds with failure load of about 7.8 kN for a 25 mm wide specimen have been reported.

In another study conducted by Wei et al. (2012), a welding tool with cutting feature pin has been designed for lap welding Mg alloy AZ31 to stainless steel SUS302. The welding tool has a tungsten carbide pin with cutting feature (rotary burr) on the surface as shown in Figure 5.40A. During welding the pin was plunged into steel and cut steel off the bulk. Figure 5.40B–D shows the transverse cross sections of welds made at 950 rpm and various traverse speeds from 60 to 300 mm/min. Lots of steel flashes and scraps are obvious at lower traverse speed with presence of void. As the traverse speed increases, less flashes and scraps can be noticed without visible void.

5.6 FSW OF DISSIMILAR MATERIALS WITH COATINGS AND ADHESIVE

Before dissimilar materials welding can be implemented for practical applications, two challenges need to be addressed. One is good weldability and repeatability with acceptable mechanical performance. The other is good resistance to galvanic corrosion. As discussed in previous sections, most of the endeavors focus on researching weldability with various metal and alloy systems, and investigation of the welding parameters, configuration influence on weld integrity, brittle IMC formation, and the relevant mechanical properties. One method of improving weldability, reducing or eliminating deleterious IMC, or forming strong but less brittle IMC, which has not been discussed in detail, is altering the mechanism of chemical reaction at weld interface. This can be done by selectively creating or placing a third material in between the two dissimilar metals or more practically applying a surface coating to at least one of the two metals. Metal or alloy coatings are generally adopted for this purpose. These coatings in some cases can improve weldability, such as galvanized steels (Zn coating on steels) which are welded to Mg alloys discussed previously, although the initial intension of Zn coating is to prevent corrosion of steel substrate. Similar observation has been made during FSW of Al alloys to Zn-coated steels as well (Chen et al., 2008), due to the formation of low-temperature Al–Zn eutectic at the weld interface, which leads to intimate contact and enhanced mutual diffusion between Al and steel. Coatings, especially organic coatings, adhesive, or sealant can also bring challenges to FSW when considered altogether dissimilar material, when corrosion resistance is the main consideration.

Figure 5.41 presents friction stir spot-welded 2.5 mm Al alloy 1050 and 1.8 mm hot-stamped boron steel with a hot dip aluminized Al–Si coating layer. The thickness of the coating is about 15–25 μm as shown in Figure 5.41A. A WC-Co tool coated with 3 μm AlCrN was

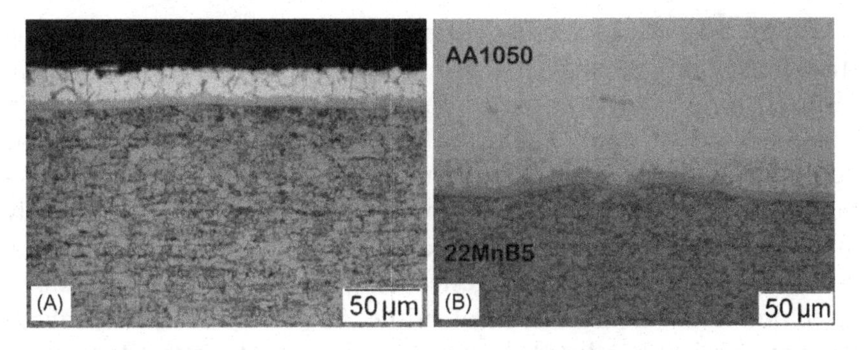

Figure 5.41 Cross section of (A) hot-stamped boron steel 22MnB5 with aluminized Al–Si surface coating and (B) friction stir spot-welded Al alloy AA1050 to coated boron steel (Da Silva et al., 2010, Reprinted with permission from Maney Publishing).

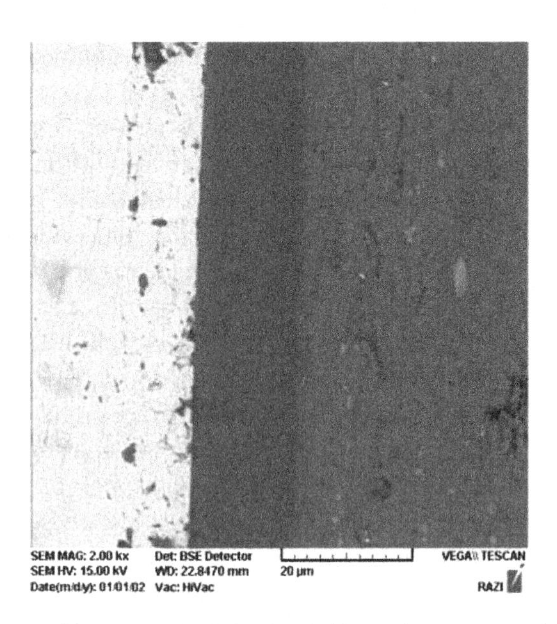

Figure 5.42 Cross section of friction stir-welded Al and Cu (Akbari et al., 2013, Reprinted with permission from Maney Publishing).

employed. It has a 12 mm diameter concave shoulder, a 2 mm conical tapered pin with pin length of 2.7 mm. A tool rotation rate of 2000 rpm and pin penetration depth of 2.9 mm were adopted to make the weld, of which the cross section is shown in Figure 5.41B. No voids or defects can be observed. It is suggested that the Al–Si coating layer appears to play an important role in the weldability of dissimilar Al and steel joints, as the interface between Al and Al–Si layer could positively enhance the diffusion ability of Al into the layer, which in turn improving the connection between both Al and steel.

Figure 5.42 shows a cross-sectional view of interface of a friction stir-welded 4 mm Al alloy AA6060 to 3 mm Cu in a lap configuration. In this case, Al alloy was placed on the top of Cu. A thin layer of Cu was anodized on the Al alloy before welding. Three zones can be noticed with the Cu layer in between of Al and Cu. Chemical composition analysis of the intermediate zone using EDS indicate a relatively uniform composition within the zone, and increased Cu content by 20% after welding. It was suggested that the use of the Cu anodized layer prevented the formation of brittle Al–Cu IMCs and enhanced the metallurgical bonding, which in turn, improved the tensile strength by 25% compared to the dissimilar welds without anodized layer.

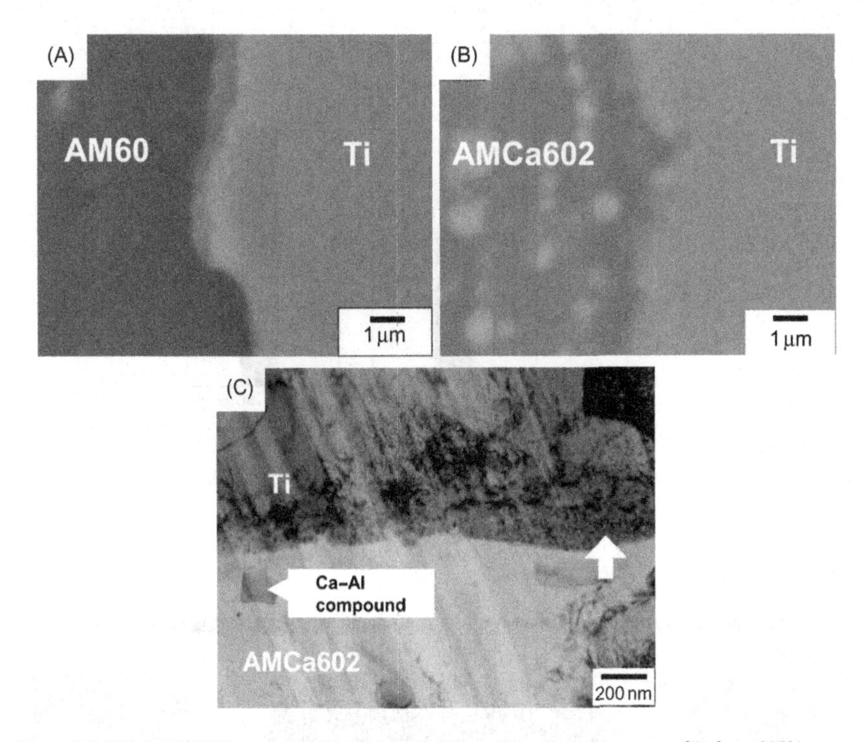

Figure 5.43 SEM and TEM images of friction stir-welded Ti and Mg alloys (Aonuma and Nakata, 2010).

Instead of adding a third material to the interface of dissimilar materials, Aonuma and Nakata (2010) studied the effect of calcium addition to an Mg alloy AM60 on IMC layer formation between dissimilar welding of Mg and Ti. An addition of 2 wt% of Ca was done to the AM60 alloy to make it a AMCa602. In this study, 2 mm AMCa602 was butt welded to a same thickness pure Ti using a tool made of alloy steel which mainly plunged into the soft Mg alloy. Figure 5.43 presents the SEM and TEM images of the weld interfaces. For dissimilar welds between AM60 and Ti (Figure 5.43A), a reaction layer is noticed at the interface. Chemical composition analysis suggested the formation of $TiAl_3$ IMC due to high Al content in Mg alloy. Figure 5.43B presents the weld interface between AMCa602 and Ti. Two kinds of layers are noticed under SEM, with one containing Mg, Ca, and Al and the other only Mg and Al according to EDS. TEM image (Figure 5.43C) shows Al_2Ca compound adjacent to the interface. It is suggested that Ca addition into Mg alloy reacted with Al and formed Al_2Ca compound and decreased the Al solid solution in the matrix of AMCa602, in turn, suppressed the formation of

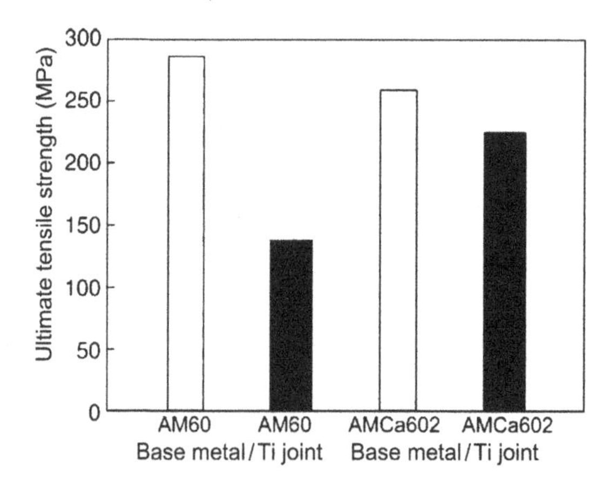

Figure 5.44 Tensile strength of friction stir-welded Mg and Ti alloys (Aonuma and Nakata, 2010).

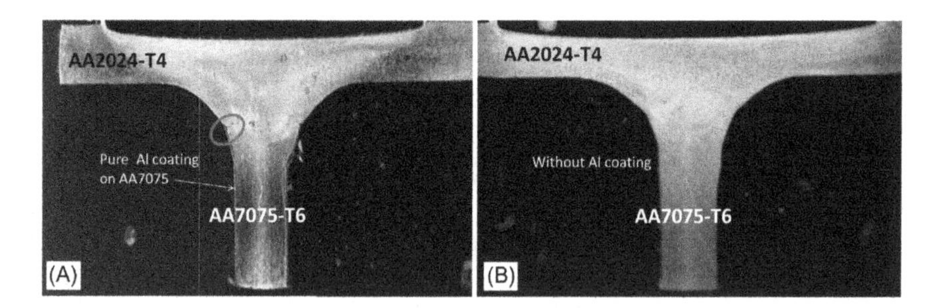

Figure 5.45 Transverse cross sections of dissimilar friction stir-welded AA2024 to (A) Al 7075 with 0.2 mm pure Al coating and (B) Al 7075 without coating (Acerra et al., 2010, Reprinted with permission from Springer).

Ti–Al IMC layer at the joint interface and improved the tensile strength significantly as shown in Figure 5.44.

Figure 5.45 presents a study carried out by Acerra et al. (2010) on FSW of dissimilar high-strength Al alloys in a T joint for aeronautical application. The Al alloys used are 2.3 mm Al 2024-T4 and 2.8 mm Al 7075-T6. The Al 7075 has a pure 0.2 mm Al surface coating for improving its corrosion resistance. During welding, the Al 2024 was selected as skin (upper material) and Al 7075 as stringer (lower material). Figure 5.45A shows a transverse cross section of a weld made at 500 rpm and 25 mm/min. A discontinuity in the weld can be noticed. Figure 5.45B shows a transverse cross section of a weld made using

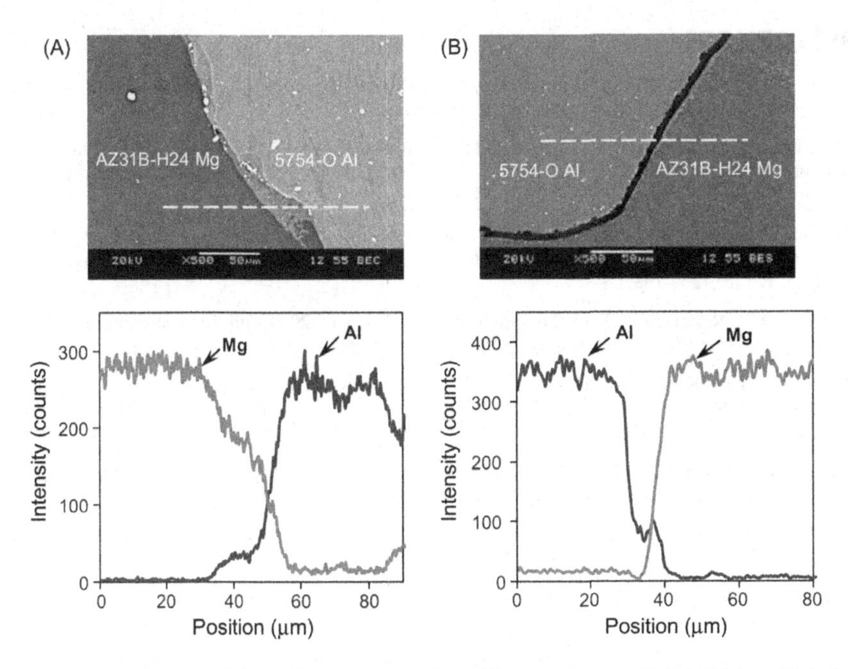

Figure 5.46 SEM images and chemical composition analysis of friction stir spot-welded AA5754 to AZ31, (A) without adhesive and (B) with adhesive at interface (Chowdhury et al., 2013).

same welding parameters and welding tool as those for Figure 5.45A, but removed surface coating. In this case, no defect or weld discontinuity can be observed. It is concluded that, the coating layer negatively interacts with material flow during dissimilar welding, which leads to macro and micro defects.

Adhesive which is used for structural application may prevent IMC formation. In a study conducted by Chowdhury et al. (2013), an epoxy-based adhesive was used in dissimilar friction stir spot welding between Al alloy 5754 and Mg alloy AZ31. The adhesive was applied to the interface between Al and Mg and hardened through curing before subjecting to welding. Figure 5.46 shows a weld interface comparison between welds with and without adhesive. For the weld without adhesive (Figure 5.46A), a thick IMC layer of thickness up to 20 μm can be clearly noticed. Composition analysis indicates mutual diffusion of Al and Mg across the layer. For the weld with hardened adhesive (Figure 5.46B), cross section of the weld shows a prominent boundary between the Al and Mg. Composition analysis indicates that adhesive at interface reduced the thickness of the IMC layer.

REFERENCES

Acerra, F., Buffa, G., Fratini, L., Troiano, G., 2010. On the FSW of AA2024-T4 and AA7075-T6 T-joints: an industrial case study. Int. J. Adv. Manuf. Technol. 48, 1149–1157.

Akbari, M., Bahemmat, P., Haghpanahi, M., Besharati Givi, M.K., 2013. Enhancing metallurgical and mechanical properties of friction stir lap welding of Al–Cu using intermediate layer. Sci. Technol. Weld. Join. 18, 518–524.

Aonuma, M., Nakata, K., 2010. Effect of calcium on intermetallic compound layer at interface of calcium added magnesium-aluminum alloy and titanium joint by friction stir welding. Mater. Sci. Eng. B 173, 135–138.

Aonuma, M., Nakata, K., 2011. Dissimilar metal joining of 2024 and 7075 aluminum alloys to titanium alloys by friction stir welding, Mater. Trans. 52, 948–952.

Bang, K.S., Lee, K.J., Bang, H.S., Bang, H.S., 2011. Interfacial microstructure and mechanical properties of dissimilar friction stir welds between 6061-T6 aluminum and Ti-6%Al-4% V alloys. Mater. Trans. 52, 974–978.

Bergmann, J.P., Schuerer, R., Ritter, K., 2013. Friction stir welding of tailored blanks of aluminum and magnesium alloys. Key Eng. Mater. 549, 492–499.

Cao, X., Jahazi, M., 2010. Friction stir welding of dissimilar AA 2024-T3 to AZ31B-H24 alloys. Mater. Sci. Forum 638–642, 3661–3666.

Chang, W.S., Kim, H.J., Kim, S.W., 2010. Microstructure and mechanical properties of dissimilar joints of AZ31 Mg alloy to aluminium alloys. Mater. Sci. Forum 638–642, 214–219.

Chen, Y.C., Nakata, K., 2008. Friction stir lap joining aluminum and magnesium alloys. Scr. Mater. 58, 433–436.

Chen, Y.C., Komazaki, T., Tsumura, T., Nakata, K., 2008. Role of zinc coat in friction stir lap welding Al and zinc coated steel. Mater. Sci. Technol. 24, 33–39.

Chowdhury, S.H., Chen, D.L., Bhole, S.D., Cao, X., Wanjara, P., 2013. Lap shear strength and fatigue behavior of friction stir spot welded dissimilar magnesium-to-aluminum joints with adhesive. Mater. Sci. Eng. A 562, 53–60.

Coelho, R.S., Kostka, A., dos Santos, J.F., Kaysser-Pyzalla, A., 2012. Friction-stir dissimilar welding of aluminium alloy to high strength steels: Mechanical properties and their relation to microstructure. Mater. Sci. Eng. A 556, 175–183.

Da Silva, A.A.M., Aldanondo, E., Alvarez, P., Arruti, E., Echeverría, A., 2010. Friction stir spot welding of AA 1050 Al alloy and hot stamped boron steel (22MnB5). Sci. Technol. Weld. Join. 15, 682–687.

Firouzdor, V., Kou, S., 2009. Al-to-Mg friction stir welding: effect of positions of Al and Mg with respect to the welding tool. Weld. J. (Miami, FL) 88, 213S–224S.

Firouzdor, V., Kou, S., 2010a. Formation of liquid and intermetallics in Al-to-Mg friction stir welding. Metall. Mater. Trans. A 41, 3238–3251.

Firouzdor, V., Kou, S., 2010b. Al-to-Mg friction stir welding: effect of material position, travel speed, and rotation speed. Metall. Mater. Trans. A 41, 2914–2935.

Galvão, I., Oliveira, J.C., Loureiro, A., Rodrigues, D.M., 2011. Formation and distribution of brittle structures in friction stir welding of aluminum and copper: influence of process parameters. Sci. Technol. Weld. Join. 16, 681–689.

Jana, S., Hovanski, Y., 2012. Fatigue behaviour of magnesium to steel dissimilar friction stir lap joints. Sci. Technol. Weld. Join. 17, 141–145.

Jana, S., Hovanski, Y., Grant, G.J., 2010. Friction stir lap welding of magnesium alloy to steel: a preliminary investigation. Metall. Mater. Trans. A 41, 3173–3182.

Khodir, S.A., Shibayanagi, T., 2007. Dissimilar friction stir welded joints between 2024-T3 aluminum alloy and AZ31 magnesium alloy. Mater. Trans. 48, 2501–2505.

Kostka, A., Coelho, R.S., dos Santos, J., Pyzalla, A.R., 2009. Microstructure of friction stir welding of aluminium alloy to magnesium alloy. Scr. Mater. 60, 953–956.

Kwon, Y.J., Shigematsu, I., Saito, N., 2008. Dissimilar friction stir welding between magnesium and aluminum alloys. Mater. Lett. 62, 3827–3829.

Liang, Z., Chen, K., Wang, X., Yao, J., Yang, Q., Zhang, L., et al., 2013. Effect of tool offset and tool rotational speed on enhancing mechanical property of Al/Mg dissimilar FSW joints. Metall. Mater. Trans. A 44, 3721–3731.

Liu, X., Lan, S., Ni, J., 2014. Analysis of process parameters effects on friction stir welding of dissimilar aluminum alloy to advanced high strength steel. Mater. Des. 59, 50–62.

Liyanage, T., Kilbourne, J., Gerlich, A.P., North, T.H., 2009. Joint formation in dissimilar Al alloy/steel and Mg alloy/steel friction stir spot welds. Sci. Technol. Weld. Join. 14, 500–508.

Malarvizhi, S., Balasubramanian, V., 2012. Influences of tool shoulder diameter to plate thickness ratio (D/T) on stir zone formation and tensile properties of friction stir welded dissimilar joints of AA6061 aluminum-AZ31B magnesium alloys. Mater. Des. 40, 453–460.

Masoudian, A., Tahaei, A., Shakiba, A., Sharifianjazi, F., Mohandesi, J.A., 2014. Microstructure and mechanical properties of friction stir weld of dissimilar AZ31-O magnesium alloy to 6061-T6 aluminum alloy. Trans. Nonferrous Met. Soc. China (English Edition) 24, 1317–1322.

Massalski, T., 1990. Binary alloy phase diagrams. ASM Int. 1, 141–143.

McLean, A.A., Powell, G.L.F., Brown, I.H., Linton, V.M., 2003. Friction stir welding of magnesium alloy AZ31B to aluminium alloy 5083. Sci. Technol. Weld. Join. 8, 462–464.

Mofid, M.A., Abdollah-Zadeh, A., Ghaini, F.M., CGür, C.H., 2012. Submerged friction-stir welding (SFSW) underwater and under liquid nitrogen: An improved method to join Al alloys to Mg alloys. Metall. Mater. Trans. A 43, 5106–5114.

Mofid, M.A., Abdollah-Zadeh, A., Gür, C.H., 2014. Investigating the formation of intermetallic compounds during friction stir welding of magnesium alloy to aluminum alloy in air and under liquid nitrogen. Int. J. Adv. Manuf. Technol. 71, 1493–1499.

Morishige, T., Kawaguchi, A., Tsujikawa, M., Hino, M., Hirata, T., Higashi, K., 2008. Dissimilar welding of Al and Mg alloys by FSW. Mater. Trans. 49, 1129–1131.

Murray, J.L., 1992. Al (Aluminum) Binary Alloy Phase Diagrams, Alloy Phase Diagrams, vol. 3. ASM Handbook, ASM International, pp. 2.4–2.56.

Regev, M., Mehtedi, M.E., Cabibbo, M., Quercetti, G., Ciccarelli, D., Spigarelli, S., 2014. High temperature plasticity of bimetallic magnesium and aluminum friction stir welded joints. Metall. Mater. Trans. A 45, 752–764.

Sato, Y.S., Park, S.H.C., Michiuchi, M., Kokawa, H., 2004. Constitutional liquation during dissimilar friction stir welding of Al and Mg alloys. Scr. Mater. 50, 1233–1236.

Schneider, C., Weinberger, T., Inoue, J., Koseki, T., Enzinger, N., 2011. Characterization of interface of steel/magnesium FSW. Sci. Technol. Weld. Join. 16, 100–106.

Shigematsu, I., Kwon, Y.J., Saito, N., 2009. Dissimilar friction stir welding for tailor-welded blanks of aluminum and magnesium alloys. Mater. Trans. 50, 197–203.

Simoncini, M., Forcellese, A., 2012. Effect of the welding parameters and tool configuration on micro- and macro-mechanical properties of similar and dissimilar FSWed joints in AA5754 and AZ31 thin sheets. Mater. Des. 41, 50–60.

Song, Z., Nakata, K., Wu, A., Liao, J., Zhou, L., 2014. Influence of probe offset distance on interfacial microstructure and mechanical properties of friction stir butt welded joint of Ti6Al4V and A6061 dissimilar alloys. Mater. Des. 57, 269–278.

Tan, C.W., Jiang, Z.G., Li, L.Q., Chen, Y.B., Chen, X.Y., 2013. Microstructural evolution and mechanical properties of dissimilar Al–Cu joints produced by friction stir welding. Mater. Des. 51, 466–473.

Tanaka, T., Morishige, T., Hirata, T., 2009. Comprehensive analysis of joint strength for dissimilar friction stir welds of mild steel to aluminum alloys. Scr. Mater. 61, 756–759.

Uzun, H., Dalle Donne, C., Argagnotto, A., Ghidini, T., Gambaro, C., 2005. Friction stir welding of dissimilar Al 6013-T4 To X5CrNi18-10 stainless steel. Mater. Des. 26, 41–46.

Watanabe, T., Takayama, H., Yanagisawa, A., 2006. Joining of aluminum alloy to steel by friction stir welding. J. Mater. Process. Technol. 178, 342–349.

Wei, Y., Li, J., Xiong, J., Huang, F., Zhang, F., 2012. Microstructures and mechanical properties of magnesium alloy and stainless steel weld-joint made by friction stir lap welding. Mater. Des. 33, 111–114.

Xue, P., Xiao, B.L., Ni, D.R., Ma, Z.Y., 2010. Enhanced mechanical properties of friction stir welded dissimilar Al–Cu joint by intermetallic compounds. Mater. Sci. Eng. A 527, 5723–5727.

Xue, P., Ni, D.R., Wang, D., Xiao, B.L., Ma, Z.Y., 2011. Effect of friction stir welding parameters on the microstructure and mechanical properties of the dissimilar Al–Cu joints. Mater. Sci. Eng. A 528, 4683–4689.

Yamamoto, N., Liao, J, Watanabe, S., Nakata, K., 2009. Effect of intermetallic compound layer on tensile strength of dissimilar friction-stir weld of a high strength Mg alloy and Al alloy. Mater. Trans. 50, 2833–2838.

Yan, J., Xu, Z., Li, Z., Li, L., Yang, S., 2005. Microstructure characteristics and performance of dissimilar welds between magnesium alloy and aluminum formed by friction stirring. Scr. Mater. 53, 585–589.

Yan, Y., Zhang, D.T., Qiu, C., Zhang, W., 2010. Dissimilar friction stir welding between 5052 aluminum alloy and AZ31 magnesium alloy. Trans. Nonferrous Met. Soc. China (English Edition) 20, s619–s623.

Zettler, R., Da Silva, A.A.M., Rodrigues, S., Blanco, A., Dos Santos, J.F., 2006. Dissimilar Al to Mg alloy friction stir welds. Adv. Eng. Mater. 8, 415–421.

CHAPTER 6

Modeling and Simulation of Friction Stir Welding of Dissimilar Alloys and Materials

A tremendous increase in computational power in the last one decade or so has propelled use of modeling and simulation in the study of many engineering problems across various disciplines. It has also accelerated the development of more efficient and powerful finite element based commercial codes such ANSYS, ABAQUS, and COMSOL for solving engineering problems. The friction stir welding (FSW) community has also benefited from such development taking place in the area of modeling and simulation.

As FSW technology is making inroads in ground, aerospace, and sea transportation industries, it becomes important to understand various aspects of FSW such as heat generation, material flow, and microstructure evolution for an improved understanding of the processing and efficient use of the process in many areas. So far the trend has been to use experimental techniques to study these aspects. However, experimental techniques are time consuming, tedious, and very costly. Moreover, experiments are generally carried out with a set of objectives in mind. For example, one particular group might be interested in studying just material flow aspect of FSW. However, simulation of FSW is capable of providing information which is much beyond the reach of individual researchers. For example, simulation of the process can provide not only information on heat generation, heat conduction, and material flow but also on residual stresses (RSs). Experimentally, the RS information is available at no more than a few locations in the entire weldment which is inadequate for developing an overall understanding of RS distribution in the weldments.

FSW is a fully coupled thermo-mechanical process which involves frictional and adiabatic heating of the workpiece, very large strain and strain rate, and very complex material flow during the welding process.

Friction Stir Welding of Dissimilar Alloys and Materials. DOI: http://dx.doi.org/10.1016/B978-0-12-802418-8.00006-0

Successful simulation of FSW process not only involves consideration of all essential elements (e.g., tool, backing plates, and clamps) but also demands for accurate material constitutive relationships, knowledge of variation of friction coefficients as a function of temperature, and condition of materials at tool–workpiece interface (stick or slip). In view of these conditions, it has been very challenging to carry out process modeling of even similar FSW processes. Despite all difficulties, however, several attempts have been made toward modeling and simulation to capture essential physics of the process (He et al., 2014).

With regard to simulation of FSW using finite element analysis (FEA), in general two different paths have been adopted by various researchers—computational fluid dynamics (CFD) based approach and continuum solid mechanics based approach. CFD based models have been used largely to predict the thermal history during the welding. The thermal history is subsequently used in a solid mechanics based model to impose a prescribed temperature history and predict the resulting RS in the weldments. Solid mechanics based models are capable of predicting both—thermal history and RS. Two different approaches exist within solid mechanics continuum based models—(i) sequentially coupled and (ii) fully coupled thermo-mechanical model (Nandan et al., 2008a).

The simulation work to understand material flow during FSW carried out by Colegrove and Shercliff (2004) is shown in Figure 6.1. In that work, two different geometries of the pin were considered: Trivex and Triflute. To simulate the material flow, a commercial code FLUENT based on CFD principle was used. Figure 6.1A shows that the streamlines, in line with the tool, are swept around the rotating tool from advancing side to rotating side in the direction of the tool rotation. The same observation can be made by looking at the movement of streamlines for the tool shown in Figure 6.1C. Figure 6.1B includes only one streamline and clearly shows an upward movement of the streamline. Such kind of movement is absent for Trivex tool shown in Figure 6.1C. The simulation output, to a large extent, is dependent on type of assumptions made during the development of the FSW process model and accuracy of thermo-mechanical properties used in the model. For example, Colegrove and Shercliff (2004) indicated that an experiment carried out using Trivex tool does show upward movement of the material.

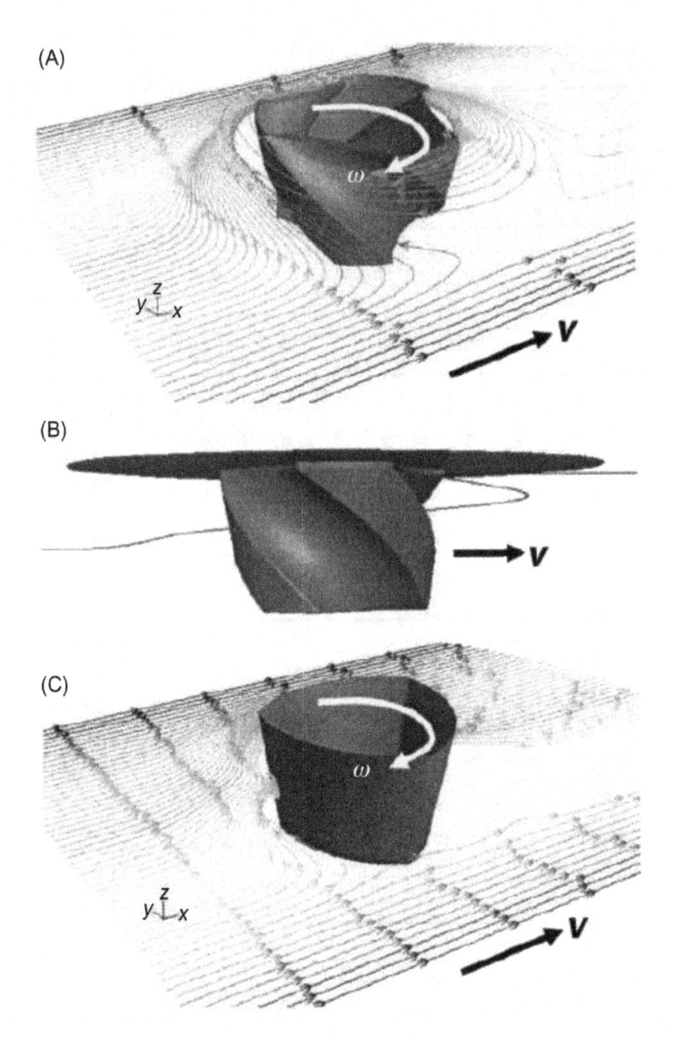

Figure 6.1 Simulation of material flow as a function of tool geometry using CFD package FLUENT (A) & (B) Triflute and (C) Trivex tools (Colegrove and Shercliff, 2004, Reprinted with permission from Maney Publishing).

A computed thermal profile for Ti–6Al–4V alloy is shown in Figure 6.2. It shows two-dimensional temperature contours at different sections of the plate during FSW. It is generally not possible to obtain thermal profiles from the nugget region where significant plastic deformation takes place. The thermocouple placed in the path of the rotating and advancing tool will either get destroyed due to shear forces or get displaced from its original position. This difficulty highlights the importance of thermal modeling which is capable of providing us with

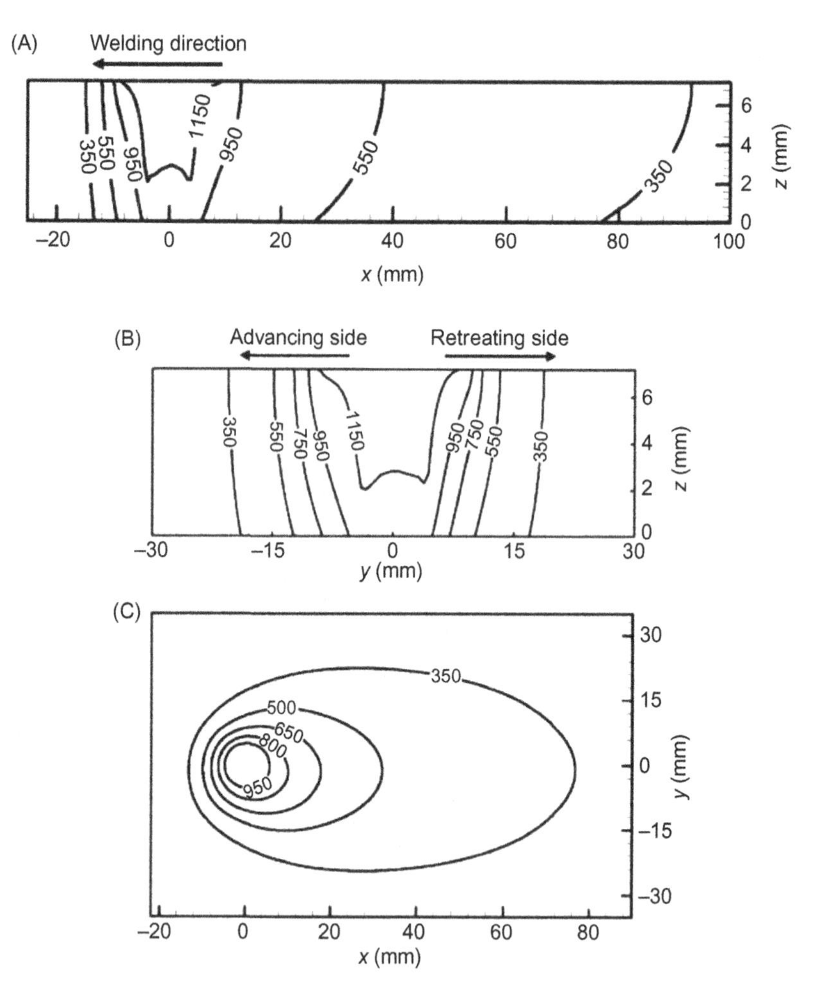

Figure 6.2 Computed thermal profile during FSW of Ti–6Al–4V alloy, (A) xz plane (welding direction), (B) yz plane (transverse cross-section), and (C) xy plane (top surface) (from Nandan et al., 2008a,b, © Carl Hanser Verlag, Muenchen).

temperature distribution even in a zone where experimental temperature measurement is not possible. The information on temperature distribution in nugget of the weld is necessary for understanding RS evolution, microstructural modeling, and mechanical properties modeling.

It is also possible to obtain information on strain and strain rate using simulation in the deformation zones of the friction stir welds. Two examples (Figures 6.1 and 6.2) presented here represent the case of similar metal welding. To turn our attention now to dissimilar

FSW, first the steady-state thermal energy conservation equation is given (Eq. (6.1)) (Nandan et al., 2006).

$$\rho C_{\mathrm{p}} \frac{\partial(u_i T)}{\partial x_i} = -\rho C_{\mathrm{p}} U \frac{\partial T}{\partial x_i} + \frac{\partial}{\partial x_i}\left(k \frac{\partial T}{\partial x_i}\right) + S_{\mathrm{b}} \quad (6.1)$$

where

ρ = density,
C_{p} = specific heat capacity,
u_i = material velocity in the direction, $i = 1,2,3$,
T = temperature,
U = welding velocity,
k = thermal conductivity,
S_{b} = heat generation rate per unit volume.

Using finite difference analysis and FEA, Eq. (6.1) can be easily solved. But even for similar welding, some of the material properties present in Eq. (6.1) are not available for all materials. Even when they are available, in most cases the temperature dependence of these material properties is unknown. In this context, the simulation of dissimilar material welding adds another layer of complexity to the existing problem. In dissimilar welding, now the models require, as input, two different types of material properties and their temperature dependence for simulating the FSW process accurately. In the case of dissimilar welding, the mixing between two different materials results in a new chemistry. For material mixing, therefore, it also becomes important to consider interdiffusivity of both materials being welded. So, the predictive capability of the simulation tool will depend on how accurately such details are captured in the modeling effort and how robust the material property database is to take care of such a complex nature of FSW process in dissimilar welding.

Due to issues highlighted above, very few simulation works exist pertaining to dissimilar FSW (Aval et al., 2012; Al-Badour et al., 2014). The model used by Aval et al. (2012) is shown in Figure 6.3. In the simulation work, temperature-dependent materials properties were used. Since FSW involves severe plastic deformation, elements based on Lagrangian formulation cannot handle such a large deformation. Hence, for such purpose, Eulerian based formulation is better. In Eulerian formulation, mesh remains stationary and material flows

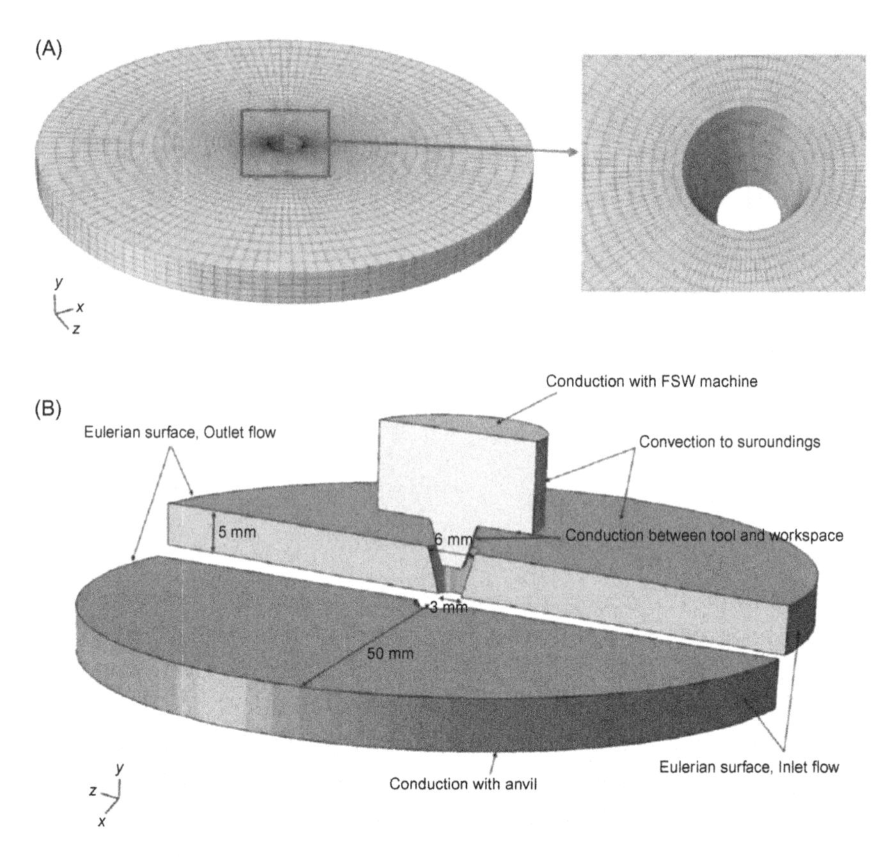

Figure 6.3 The FSW model used in the simulation of dissimilar welding of AA5086-O and AA6061-T6 alloys (Aval et al., 2012, Reprinted with permission from Springer).

through it. Hence, the issue of severe distortion of the mesh due to plastic deformation during FSW does not come into the picture. The merits and demerits of this approach have been discussed by Michaleris (2011). This approach (Eulerian formulation) was used by Aval et al. (2012) in the simulation of dissimilar FSW process. The evolution of the thermal field based on their study is shown in Figure 6.4. It shows temperature fields at 5, 10, 15, and 20 s. With increasing time, an expanding thermal field should be noted. At 15 s a steady state is reached. Careful observation of the thermal field reveals that the temperature distribution in the welded sheets is asymmetric. Note that AA6061-T6 was placed on the retreating side of the weld. Experimental measurement of temperature also confirms the existence of asymmetry of temperature distribution in the sheets during the welding.

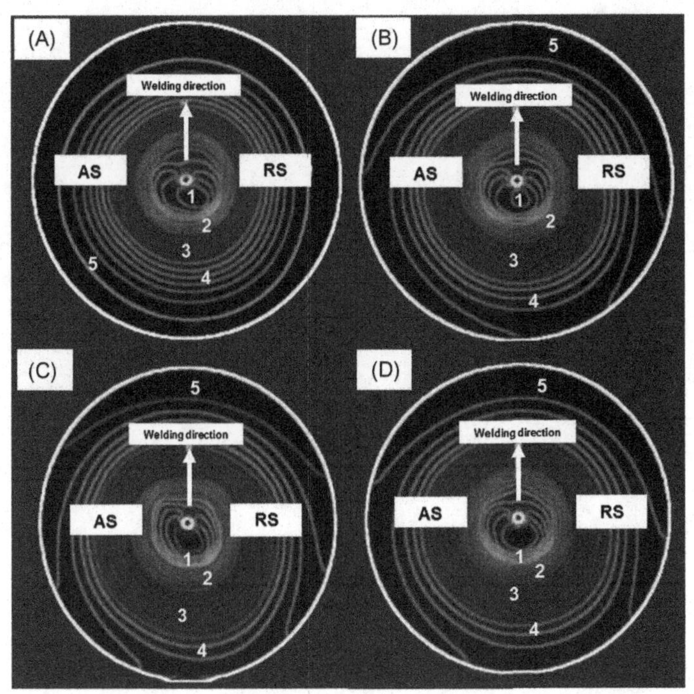

Figure 6.4 The FSW model used in the simulation of dissimilar welding of AA5086-O and AA6061-T6 alloys (A) 5 s, (B) 10 s, (C) 15 s, and (D) 20 s (Adapted from Aval et al., 2012).

REFERENCES

Al-Badour, F., Merah, N., Shuaib, A., Bazoune, A., 2014. Thermo-mechanical finite element model of friction stir welding of dissimilar alloys. Int. J. Adv. Manuf. Technol. 72, 607–617.

Aval, H.J., Serajzadeh, S., Kokabi, A.H., 2012. Experimental and theoretical evaluations of thermal histories and residual stresses in dissimilar friction stir welding of AA5086-AA6061. Int. J. Adv. Manuf. Technol. 61, 149–160.

Colegrove, P.A., Shercliff, H.R., 2004. Development of Trivex friction stir welding tool Part 2—three-dimensional flow modelling. Sci. Technol. Weld. Join. 9, 352–361.

He, X., Gu, F., Ball, A., 2014. A review of numerical analysis of friction stir welding. Prog. Mater. Sci. 65, 1–66.

Michaleris, P., 2011. Modelling welding residual stress and distortion: current and future research trends. Sci. Technol. Weld. Join. 16, 363–368.

Nandan, R., Roy, G.G., Debroy, T., 2006. Numerical simulation of three-dimensional heat transfer and plastic flow during friction stir welding. Metall. Mater. Trans. A 37, 1247–1259.

Nandan, R., DebRoy, T., Bhadeshia, H.K.D.H., 2008a. Recent advances in friction-stir welding—process, weldment structure and properties. Prog. Mater. Sci. 53, 980–1023.

Nandan, R., Lienert, T.J., DebRoy, T., 2008b. Toward reliable calculations of heat and plastic flow during friction stir welding of Ti–6Al–4V alloy. Int. J. Mater. Res. 99, 434–444.

CHAPTER 7

Challenges and Opportunities for Friction Stir Welding of Dissimilar Alloys and Materials

Friction stir welding (FSW) has shown great potential in joining various dissimilar alloy or material systems with quite different physical or chemical properties, which could not be possible for a conventional welding technique. Industrial implementation of friction stir-welded dissimilar materials, especially for the automotive industry, has displayed considerable benefits in terms of design flexibility and weight reduction for overall structures. Before FSW can be implemented extensively for dissimilar materials in wide ranging industries, the following challenges need to be addressed, so robust and reliable welds with good corrosion resistance can be achieved.

7.1 FORMATION OF DETRIMENTAL INTERMETALLIC COMPOUNDS

The formation of intermetallic compounds (IMCs) has been seen in most of the dissimilar materials FSW with different base metals, such as Al to steel, Mg or Cu. The formation and growth of these IMCs are very rapid as a result of enhanced diffusion during welding. So far, the formation of IMC is considered not avoidable in general, though the thickness of IMC can be reduced in some cases by reducing the frictional heat input or using a third media to quickly take away the heat. It has been testified that the reduction in thickness of IMC can improve the weld strength significantly. On the other hand, formation of preferable IMC which is strong but less brittle through alloying or introducing an intermedia coating or transition layer can also contribute to strength improvement. Attention should be paid in future to understand IMC formation and growth mechanisms, and explore alloying addition that can retard diffusion pathways or create less harmful IMC. Coatings and adhesive which can reduce the mutual

Friction Stir Welding of Dissimilar Alloys and Materials. DOI: http://dx.doi.org/10.1016/B978-0-12-802418-8.00007-2

diffusion process but not introduce any critical challenge to weld integrity may be also researched. Material flow and mixing through process development, particularly by tool design, play an important role in minimizing deleterious IMC.

7.2 INCIPIENT MELTING AND SOLIDIFICATION STRUCTURE

Although FSW does not involve bulk melting, it has been seen, when dissimilar materials, such as Al to Mg alloys, Al or Mg alloys to Zn-coated steels, are welded, localized melting due to low-temperature eutectic reaction can be expected. Solidification microstructure with porosities and cracks has been often noticed accompanying the formation of IMC. Although the solidification microstructure is less detrimental to the strength of welds compared to the impact that thicker IMC has, there is a lack of information on its influence on fatigue performance. Cracks or porosities inside of welds can reduce the fatigue life tremendously, so future efforts are needed to prevent local melting by reducing heat input or better control of local chemical gradients. Tool designs are expected to have more important influence on such dissimilar materials welding than welding of same type of material. Tool design, for example, stationary tool shoulder design, which introduces uniform temperature distribution in the weld, but significantly reduced frictional heat input, may be beneficial for dissimilar materials FSW.

7.3 RELIABILITY AND DURABILITY

One of the critical challenges to bring friction stir-welded dissimilar materials to practical application is reliability and durability. Before industries have enough confidence on the material combination itself, the repeatability of mass production of these dissimilar welds, and the long-term performance of the welds, it is difficult to move this exciting technique from lab bench to broad industrial applications. Most of the existing research focuses on exploring the influence of welding conditions on weldability of dissimilar welds and their static performance; and very limited attention has been given to repeatability of the welds and the fatigue performance. Unlike FSW of similar materials, the complexity of dissimilar materials makes it highly possible that weld may have different properties from one to another, or even in one weld due to welding process fluctuation. IMC or welding defects such as porosity or crack may only occur at center locations in a weld. Thus, the fatigue performance of welds needs to be fully explored. It weighs more than the static strength in terms of structural design.

7.4 CORROSION, GALVANIC CORROSION, AND STRESS CORROSION CRACKING

Corrosion, galvanic corrosion, and stress corrosion cracking are areas that have already shown impacts on structural integrity and service life. They place another challenge to bring dissimilar material welds into industrial applications, although the majority of the metallic components or structures exposed to open environment are protected against corrosion, such as painting or coating. It will be important to understand how the FSW process affects the corrosion property of individual material and the galvanic corrosion of the dissimilar weld structure, since it can accelerate the degradation process once the protection coating is worn out. FSW introduces less residual stress in a weld compared to conventional fusion welding processes. The residual stress in a dissimilar weld may be more complex especially when the two materials have very different melting temperatures and thermal conductivities. It is necessary to evaluate the stress corrosion cracking behavior as well, particularly when the dissimilar welds undergo thermo-mechanical fatigue during service.

7.5 TOOL WEAR

Tool wear is critical when dissimilar material welding involves at least one high-melting-temperature material. Although it has been shown that sound dissimilar welds may be achieved without plunging the welding tool into hard material, it is generally accepted that active material mixing between dissimilar materials via pin stirring and scribing is required for considerable mechanical performance by disrupting the tenacious oxide layer present on the surfaces, preventing IMC thickening, and forming either a metallurgical bonding or a mechanical interlocking. So tool wear has to be considered when high-melting-temperature material is adopted for the dissimilar welding, coating technology that can reduce tool wear should also be studied.

7.6 INADEQUATE MATERIAL MIXING BETWEEN SOFTER AND HARDER MATERIALS

As mentioned before, a proper material mixing between dissimilar materials is required from mechanical performance point of view. As a result of different physical properties of materials and asymmetric heat input between advancing and retreating sides of the weld, material mixing can be process and workpiece placement dependent. It becomes even more challenging when a softer material and a much harder material are friction stir welded. There have been reports on hybrid dissimilar welding of dissimilar materials by incorporating additional heat source, such as laser and arc, to soften the hard material, so material mixing can be improved with less tool wear. The approach is of practical interest to industrial application, more efforts may be needed to fully explore the capacity.

7.7 OPPORTUNITY: AEROSPACE, AUTOMOTIVE, MARINE, AND ENERGY

Although many challenges need to be addressed before FSW can be fully implemented for welding dissimilar materials in broader industry sectors, it brings opportunities and hope to industry that welding of conventionally non-weldable dissimilar materials is possible; just as the invention of FSW technology impacted the Al alloy welding industry. Aerospace, marine, and transportation industries which need high-strength lightweight structures are likely to benefit from dissimilar FSW. FSW of dissimilar high-strength high-corrosion-resistant Al alloys could give more flexibility for design of aerospace and ship structures. FSW of dissimilar Al alloys to steels has already been implemented for mass production of automobile structural components. Further expanding it to dissimilar welding of Al and Mg alloys, or Mg alloys and steels could further realize weight reduction. Dissimilar welding between dissimilar steels, steel and superalloys may open up new opportunity for nuclear and power plant industries.

Printed in the United States
By Bookmasters